World Regional Geography Book Series

Series Editor
E.F.J. de Mulder

What does Finland mean to a Finn, Sichuan to a Sichuanian, and California to a Californian? How are physical and human geographical factors reflected in their present-day inhabitants? And how are these factors interrelated? How does history, culture, socio-economy, language and demography impact and characterize and identify an average person in such regions today? How does that determine her or his well-being, behaviour, ambitions and perspectives for the future? These are the type of questions that are central to The World Regional Geography Book Series, where physically and socially coherent regions are being characterized by their roots and future perspectives described through a wide variety of scientific disciplines. The Book Series presents a dynamic overall and in-depth picture of specific regions and their people. In times of globalization renewed interest emerges for the region as an entity, its people, its landscapes and their roots. Books in this Series will also provide insight in how people from different regions in the world will anticipate on and adapt to global challenges as climate change and to supra-regional mitigation measures. This, in turn, will contribute to the ambitions of the International Year of Global Understanding to link the local with the global, to be proclaimed by the United Nations as a UN-Year for 2016, as initiated by the International Geographical Union. Submissions to the Book Series are also invited on the theme 'The Geography of…', with a relevant subtitle of the authors/editors choice. Proposals for the series will be considered by the Series Editor and International Editorial Board. An author/editor questionnaire and instructions for authors can be obtained from the Publisher, Dr. Michael Leuchner (michael.leuchner@springer.com).

More information about this series at http://www.springer.com/series/13179

Walter M. Goldberg

The Geography, Nature and History of the Tropical Pacific and its Islands

 Springer

Walter M. Goldberg
Department of Biological Sciences
Florida International University
Miami, FL, USA

ISSN 2363-9083   ISSN 2363-9091   (electronic)
World Regional Geography Book Series
ISBN 978-3-319-88795-1   ISBN 978-3-319-69532-7   (eBook)
https://doi.org/10.1007/978-3-319-69532-7

Printed on acid-free paper

This Springer imprint is published by Springer Nature
The registered company is Springer International Publishing AG
The registered company address is: Gewerbestrasse 11, 6330 Cham, Switzerland

Preface

The tropical Pacific represents the largest oceanic expanse on Earth. From the westernmost point of South America to western New Guinea, and from the Tropic of Cancer to the Tropic of Capricorn, it covers an area of more than 80 million square kilometers. In addition, archipelagoes including northwest Hawaii, the Austral Islands of French Polynesia, the Pitcairn group and Easter Island extend beyond those boundaries. Millions of people speaking different languages inhabit the thousands of high and low isles scattered within this region. Those places and their people are the focal points of this book and represent my interests in the geography of the tropical Pacific both above and below water. I wanted to write a summary of the history, culture and the changes that came to these islands in a survey that is palatable to non-specialist undergraduate students, and to those who may be simply interested in reading about places they have heard of but know little about. I weave the sciences into the other disciplines that are found within these pages wherever they warrant doing so, and this is one of the distinctions of this book. While I continue to teach scientific writing, I have purposefully written in a less formal, lighter style with minimal jargon. This book is meant to appeal to a broad audience as a primer on the region.

The introductory chapter distinguishes types of islands, and introduces their climate and geographic positions as well as the types of coral reefs associated with them. Subsequent chapters are structured by time. What is known of the initial colonists and how they migrated to New Guinea tens of thousands of years ago is described, along with the Lapita and other migrants about 3500 years ago that gave rise to the cultural divisions recognized today. From that point the influence of European explorers is detailed beginning with the Portuguese and Spanish in the sixteenth century, and later by contact by the British and French. An emphasis is placed on the central role the tropical islands of the Pacific have played through cultural exchange and trade with Europe, the United States and China. The introduction of non-native species is also stressed. As traders, whalers and others

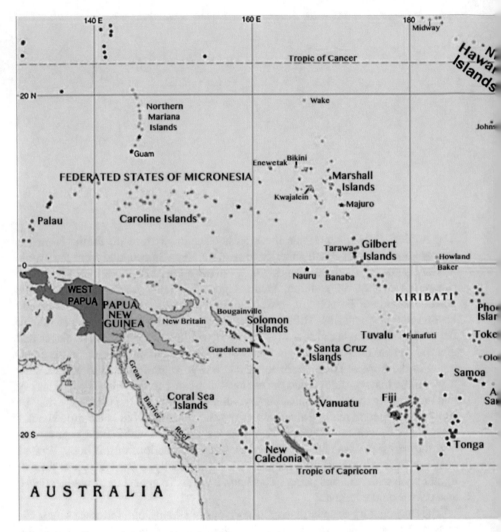

Fig. 1.0   Islands of the tropical Pacific

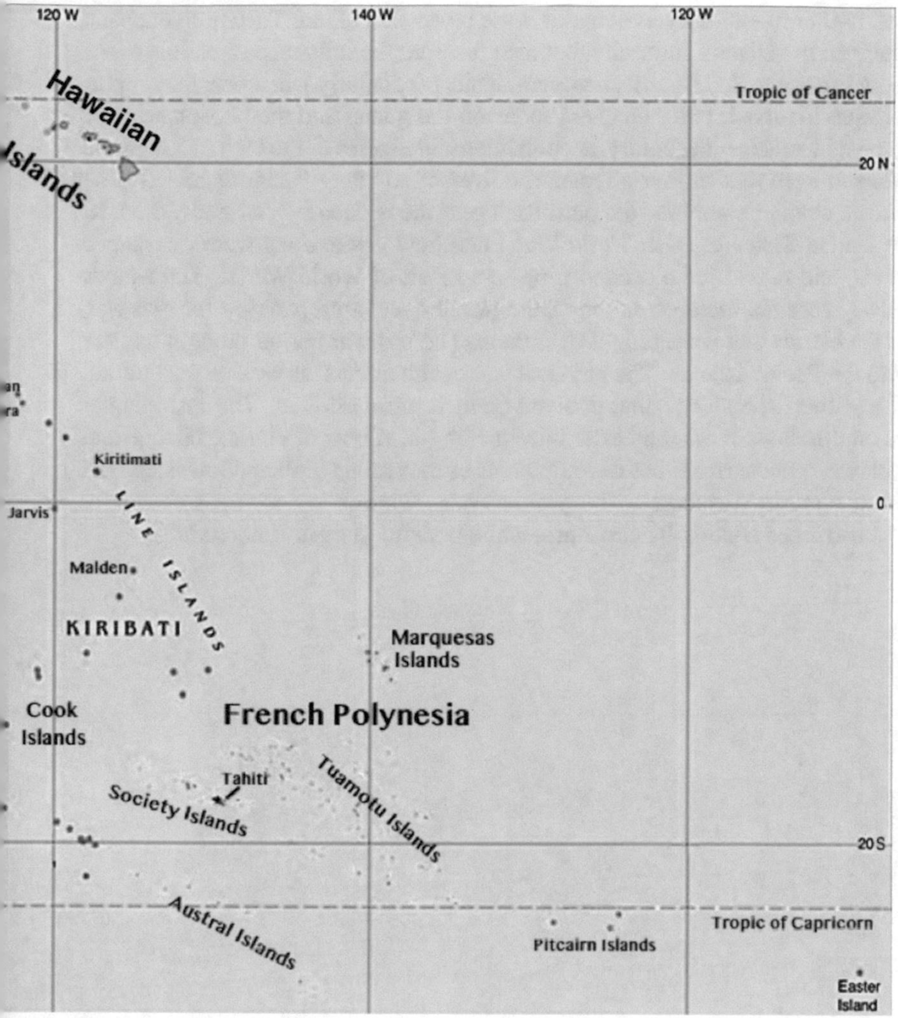
120 W
140 W
120 W
Tropic of Cancer
Hawaiian Islands
20 N
Kiritimati
Jarvis
0
LINE ISLANDS
Malden
KIRIBATI
Marquesas Islands
Cook Islands
French Polynesia
Tahiti
Society Islands
Tuamotu Islands
20 S
Austral Islands
Pitcairn Islands
Tropic of Capricorn
Easter Island

explored farther afield, the missionaries were never far behind. Their influence not only changed the islands culturally, but also fostered the influence of claims to the islands by European and American governments, particularly when business opportunities were involved. Later chapters focus on the guano and the phosphate trade during the mid-nineteenth century, and the intertwined roles of the United Kingdom, America and Peru that further affected the lives of so many Pacific islanders. The penultimate chapter examines the partitioning of the islands by Germany, the UK, and the United States prior to World War I and how possessions were rearranged afterwards, and reassorted a second time as a result of World War II. The islands themselves were restructured as they were paved over in preparation for that war, and for the battles that were fought over them. The postwar period brought nuclear testing to the Pacific islands. The physical and health effects, as well as the cultural impact and the dependency that evolved from it are described. The last chapter focuses on the modern era, and after introducing the effects of climate change and sea-level rise, emphasizes those developments as they affect high and low islands as well as their marine resources. The past has been witness to enormous change in these palm-fringed islands. Their future, while hopeful, remains uncertain.

Contents

Chapter 1
An Introduction to the Tropical Pacific and Types of Pacific Islands

Abstract The Pacific accounts for about 46% of the Earth's water surface and about one-third of its total surface area, making it larger than all of Earth's land area combined. The tropical portions are dotted with thousands of islands that despite their geographic position differ considerably climatically. A west-to-east decrease in rainfall is particularly notable during a typical year, amidst seasonal shifts in the trade wind-driven Intertropical Convergence Zones. Additional changes are super-imposed by the El Niño Southern Oscillation, which has the effect of moving the rainfall and warm water from west to east, periodically imposing disruptive floods in relatively dry climates, and droughts in areas that are normally wet. Likewise, occasional but destructive cyclonic storm tracks are common north and south of the equator. Islands in the tropical northwestern Pacific are particularly vulnerable to typhoons. High volcanic islands exhibit a distinct climate compared with low coral-line islands and this is reflected in the vegetation found on each. Coral reefs are typi-cal of most tropical islands in the Pacific, but they differ in form that includes fringing, barrier and atoll reefs. A brief description of geological processes that control reef formation and reef islands is also given.

Exotic, fabled and layered with history, the thousands of islands that dot the vast expanse of the tropical Pacific are low coral islands, steep volcanic ones, or some-thing in between. Those within the tropical realm, whether continental or true oce-anic islands, have had a very long history of immigration and exploration by distinctive native cultural groups, including some that have become spread over vast distances. However, that history has been indelibly altered by European and American influence during the last few centuries, followed by more recent issues of overfishing, pollution and global warming. This book provides a tour of the past, present and future of the unique features of the tropical Pacific, and focuses on island geography, politics, natural history, geology, anthropology and culture.

© Springer International Publishing AG 2018
W.M. Goldberg, *The Geography, Nature and History of the Tropical Pacific and its Islands*, World Regional Geography Book Series,
https://doi.org/10.1007/978-3-319-69532-7_1

1.1   The Tropical Pacific Ocean

The scope of the Pacific Ocean is enormous. Excluding Indonesian waters, the Pacific extends from the western portion of New Guinea all the way to South America, a distance of more than 16,000 km when measured just below the equator. With an area of 165 million km² the Pacific is by far the largest ocean on Earth. In fact, it accounts for about 46% of the Earth's water surface and about one-third of its total surface area, making it larger than all of Earth's land area combined. The Pacific is also the deepest ocean with an average depth of more than 4 km, a figure no doubt abetted by numerous gashes in the oceanic crust, the deep-sea trenches. The Pacific contains eight of the ten deepest trenches on Earth including the deepest, the Mariana Trench, which descends almost 11 km beneath the surface. The tropical Pacific is by definition focused on the equatorial regions that extend to the Tropic of Cancer 23.5° north of the equator as well as 23.5° south of the equator to the Tropic of Capricorn. While most tropical flora and fauna are found within these regions, variations in ocean *climate* (meaning long-term weather patterns) do not correspond exactly to the geographic tropics and allow them to extend a few hundred km beyond these boundaries, closer to 30° north and south of the equator.

Rainfall in the Tropical Pacific

In addition to altitude (described below), one of the most important factors in the natural history of tropical Pacific islands is the amount and distribution of rainfall. The most intense rains center on a narrow belt within the tropics where the counterclockwise circulation of the trade winds north of the equator meets the clockwise circulation of the trade winds from the southern hemisphere (Fig. 1.1). This collision forms a rain and thunderstorm belt that occurs on average 6–7° north of the equator, due in part to the effect of cold currents that head north from the Antarctic region. This rainiest region of the tropics is the Intertropical Convergence Zone (ITCZ), the location and intensity of which varies seasonally and displays other patterns as shown in Fig. 1.2.

The Pacific ITCZ is broadest and most consistent over the Islands of New Guinea, the Solomon the Caroline, and the southern Marshall Islands where 4–5 m of rainfall per year at sea level is common. Because of the position of maximum convective heating, there is typically a great deal of difference between rainfall in the southern and northern Marshall island group (Fig. 1.1). Likewise, the positions of the Line Islands cause disparity in rainfall totals, depending on their distance from the equator. For example Palmyra Atoll at 6° north of the equator receives an average of approximately 4400 mm per year, whereas Caroline Atoll at 10° south lies within a region of highly variable precipitation averaging 1500 mm annually.

Another region of high rainfall is in the southwest Pacific and extends from New Guinea to Vanuatu, Fiji and French Polynesia as an arm referred to as the South Pacific Convergence Zone (SPCZ) (Figs. 1.1 and 1.2b, d, e). This branch becomes

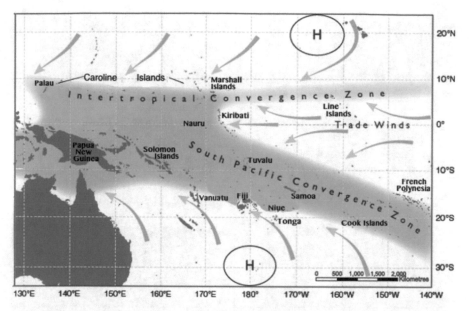

Fig. 1.1 Trade wind systems in the northern and southern hemispheres collide producing a two-armed convergence zone as shown. The northeasterly trade winds (*top arrows*) move in a clockwise direction, whereas the southwesterly trade winds (*bottom arrows*) move in the opposite direction. The most consistent annual rainfall occurs near New Guinea and the Solomon Islands. However, the arms of the system do not remain in this position [see Fig. 1.2], converging and moving seasonally to produce a dry season for islands extending toward the east. Likewise, islands that are north and south of the movement of the convergence zones experience lesser amounts of rainfall (Adapted from CSIRO and The Australian Bureau of Meteorology [1])

well developed as it moves south during the austral summer in December and January, meaning that the most intense rainfall occurs during that time. In the austral winter (June and July) the SPCZ weakens and moves north, producing a dry season in South Pacific Vanuatu, Fiji and French Polynesia.

El Niño and la Niña and their Effects on Pacific Island Climate

The normal pattern of rainfall as described above results in a western Pacific pool of warm water that is driven westward by the trade wind pattern and are 'piled up' toward New Guinea and the Solomon Islands. That effect occurs because warm water is less dense and occupies more volume than cooler water does, so that sea level around this Western Pacific Warm Pool, as it is called, is higher than the surrounding portions of the ocean (Fig. 1.1). However, at intervals of roughly 4–7 years the trade wind system weakens and allows the warm water of the western Pacific to start moving eastward along the equator (Fig. 1.3).

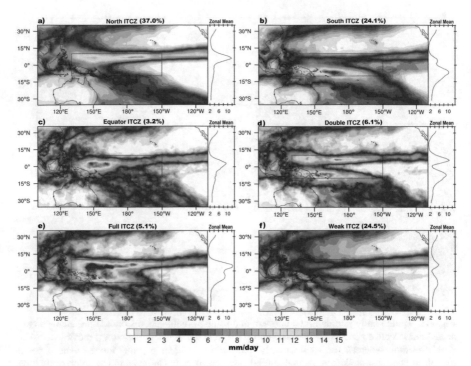

Fig. 1.2 Movement of the Intertropical Convergence Zone. The ITCZ marks the highest rainfall regions due to the intensity of heating and vertical mixing (convection) of warm moist air across the central and western Pacific. Convective heating takes its most frequent position north of the equator (**a**) in the northern summer, and the southernmost position (**b**) in the austral summer, following the apparent seasonal motion of the Sun. It may merge into a single large ITCZ on occasions when heating areas are concentrated on the equator (**c**), split in two when convection is high on either side of the equator (**d**), or spread over the equator as well as to the north and south of it (**e**). Note that c-e constitutes less than 15% of ITCZ condition frequencies. The ITCZ may also weaken as it does due to periodic wind changes and clear-sky conditions that can last for several days or two weeks at a time (**f**) (Courtesy of Baode Chen, Shanghai Typhoon Institute, China [2])

This is the beginning phase of El Niño (Spanish, 'the boy' referring to the birth of Jesus), which typically starts in December and persists into the following year (or years). El Niño affects weather patterns globally but in the Pacific, a trail of warm water may push all the way east to South America (Fig. 1.3). This causes normally dry and arid areas to become very wet and conversely, areas of typically high rainfall in the western Pacific become abnormally dry. A reversal often accompanies the end of El Niño and results in abnormally cool and dry equatorial weather called La Niña ('The girl'). The entire cycle is referred to as ENSO, El Niño Southern Oscillation (Fig. 1.4), and it can have a dramatic effect on rainfall and storms. Islands on the equator tend to have higher rainfall than normal, whereas those on either side of the equator tend to be drier. In either case crop failures may occur,

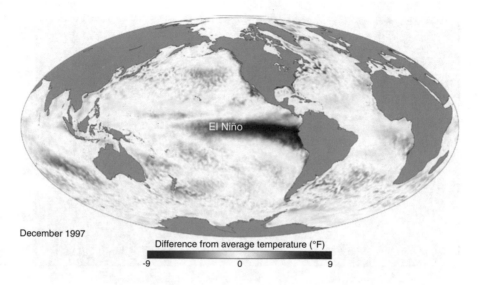

Fig. 1.3 The 1997 El Niño showing the movement of warm water from the western to the central and eastern Pacific (Courtesy of NOAA climate.gov)

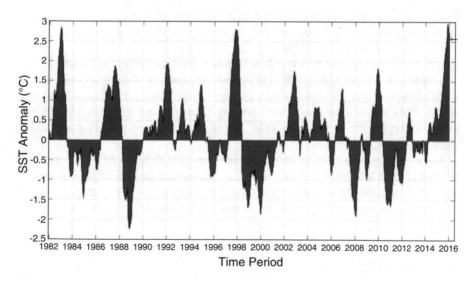

Fig. 1.4 A timeline of El Niño Southern Oscillation (ENSO) 1982–2016. *Red* indicates warm-water periods with extremes shown in 1982, 1998 and 2016. *Blue* indicates corresponding cool-water La Niña phases (Courtesy of NOAA.gov)

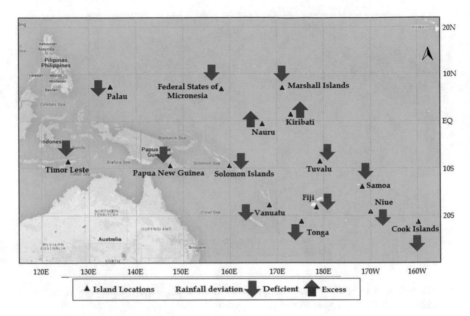

Fig. 1.5 Rainfall increases in island groups along the equator and decreases to the north and south. After United Nations ESCAP Program Report, 2014–2015 [3b]

fresh water supplies may diminish and storm frequencies may be altered as a result. Abnormally warm water also has a direct effect on coral reefs and sea-level rise, matters that will be discussed later.

In El Niño years, wet season rainfall tends to increase along the equatorial belt and on either side of it (Fig. 1.5). On islands that cover a wide geographical area, such as Papua New Guinea, Tuvalu, and the Cook Islands, El Niño effects vary across different regions within each country. The northern parts of these islands generally experience increased rainfall during an El Niño year, while the southern parts experience a decrease [3a].

For example, on the island of Tarawa in Kiribati (1.4° north of the equator) the normal rainfall is about 2000 mm per year. However, during El Niño periods that amount often more than doubles, and during La Niña phase it may drop to as little as 150 mm per year, causing declines in fresh water supplies and crop losses.

Storms and Cyclones

Hurricanes in the Atlantic are referred to as cyclones in most of the Pacific and as typhoons in the storm-prone northwestern Pacific (coastal China, the Philippines and Japan, and the Caroline and Mariana islands- see Fig. 1.7) but they are all the same. They refer to a closed, rotating system of thunderstorms that originates in or near the tropics and has a low-level circulation with a wind speed of 120 km per

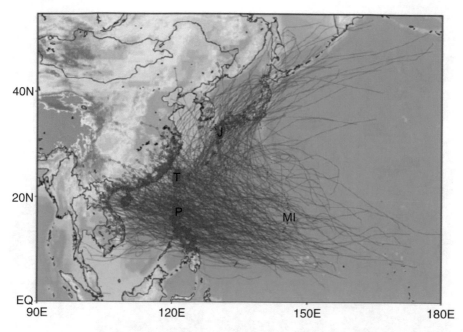

Fig. 1.7 Western Pacific typhoon-prone areas include the Mariana Islands (MI), most of the Philippines (P), southern Japan (J), Taiwan (T) and the southeast Asian mainland from Vietnam to China, all shown in *red* (Graphic courtesy of J. Winkle et al. (2012) J Climate 25:4729–4735)

hour or more. Cyclones bring destructive winds, torrential rains and storm surges. They can form massive waves that damage soil, wipe out crops for years and contaminate fresh water with seawater. Cyclones can originate anywhere in the tropical Pacific but they are typically absent from the equator ±3–5° of latitude where hot air rises and winds are typically weak. This is clearly shown in Fig. 1.6. Tropical storms gain momentum as they leave the equator (or if they form outside of it), and begin turning clockwise in the northern hemisphere, and counterclockwise in the south.

This deflection is due to the Coriolis force, an acceleration imparted by a rotating and curved Earth surface. The force is strongest toward the poles and weakest at the equator, so that large-scale rotation that characterizes cyclonic storms begins outside of the equatorial region. Equatorial islands such as the Gilbert and Phoenix groups (see Fig. 1.15) are in this rising air zone. However, most other islands of the tropical Pacific are frequently in the path of cyclones, and because of the strong tropical currents that flow north of the equator toward the north pole, the most frequent and most violent cyclones (category 5 super typhoons 253 km per hour or more) are found in the Western Pacific (Fig. 1.7). By contrast, the Peru Current originates from the Antarctic region and carries cool water toward the tropics in the eastern Pacific. This renders the tropical regions of Ecuador and Peru cyclone free.

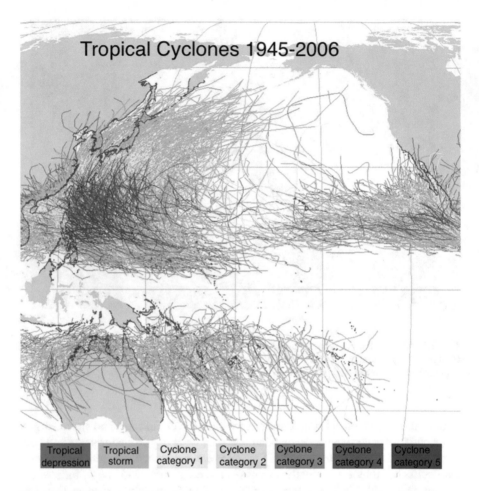

Fig. 1.6 Tropical Storm and Cyclone Tracks, 1945–2000 in the Pacific Ocean, courtesy of NOAA and Joint Typhoon Warning Center. The equator is an area that is typically free of cyclones

1.2   The Tropical Pacific: How Many and What Kind of Islands

Within the confines of the Pacific considered here (which excludes Australia, New Zealand and Japan), there are thousands of islands that fall into numerous geographic and political districts. Counting them is no easy task due to the lack of a uniform database and the necessity of relying on information sources that may be incomplete. In some cases the total number of islands may include small islets or rocks, or only larger islands, or only inhabited islands. Taking the information at face value, more than 2600 islands are included in Table 1.1. Atolls, a particular type of low island described below, may constitute the majority of islands in a particular group (whether populated or not), and this is indicated in the Table where the

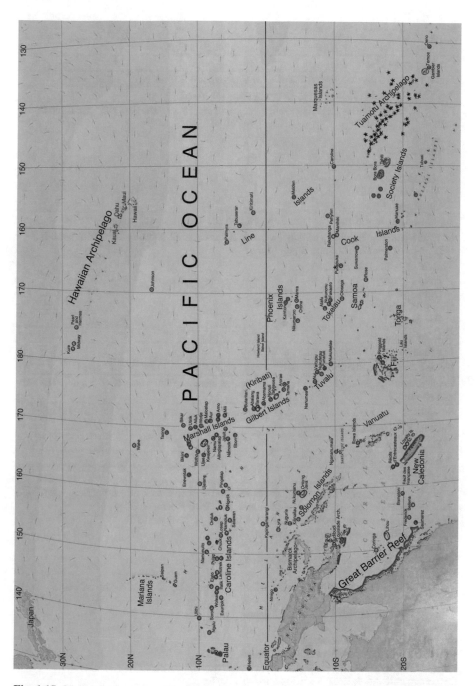

Fig. 1.15 Main island groups of the Pacific Ocean are labeled. Atolls are depicted with circles; The Tuamotu Archipelago (Tuamotu means 'cloud of islands') is the largest concentration of atolls in the world (71) is shown as stars at lower left. Barrier reefs are shown in *green*; fringing reefs are *red*. Note that atolls often occur in clusters and not all of them can be labeled on a single map. However, a more detailed and interactive map of atolls can be found at [12] where 439 have been documented, more than 80% of which are in the Pacific Ocean and adjacent seas [13] (Map after Goldberg [15])

Table 1.1 Tropical Pacific Islands

Caroline Islands: Of 607 islands, 22 main islands and 34 atolls[a]
Cook Islands 15 incl. 7 atolls[b]
Eastern Pacific 5 islands: Easter, Pitcairn, Henderson, and 2 atolls
Fiji: Of 332 islands 106 inhabited; 25 atolls, some inhabited[c]

French Pacific territories including French Polynesia and New Caledonia, 125 islands total including 84 atolls[d]
  Society 13
  Tuamotu 83
  Austral 8
  Marquesas 13
  Wallis/Fortuna 2
  Loyalty/ New Caledonia 6

Hawaiian islands to Kure: Of 137 total, 8 main islands and 10 small including 5 atolls[e]
Isolated islands:8 including Howland, Baker and Jarvis; atolls: Johnston, Palmyra, Wake (all US), and independent Niue and Nauru islands

Kiribati of 34 total, 26 are atolls[f]
  Phoenix group 8
  Gilbert group 16
  Line Islands 9
  Banaba 1

Mariana Islands 14[g]
Marshall Islands 34 islands total; incl. 29 atolls[f]
Papua New Guinea and Bismarck Archipelago: Of 151: 22 > 100 km^2 and 19 atolls[h]
Samoa/American Samoa 22 islands (8 inhabited); American Samoa includes 4 inhabited islands and two atolls[i]
Solomon/Santa Cruz Islands: Of >900 islands incl. 6 main islands, >300 inhabited; 9 atolls[d, j]
Vanuatu 83 islands (70 inhabited)[f]
Tokelau 3 (all atolls)[d]
Tonga 169 (36 inhabited); two atolls[d, k]
Tuvalu 9 (8 atolls)[l]

[a]https://en.wikipedia.org/wiki/list_of_islands_of_the-Fedrated-States_of_Micronesia encyclopedia; CIA World Fact Book 2014; all atoll numbers are from Goldberg (2016) [d]
[b]https://en.wikipedia.org/wiki/Cook_Islands; CIA World Fact Book, 2014
[c]https://en.wikipedia.org/wiki/Geography_of_Fiji
[d]Goldberg 2016 Atoll Res Bull 610
[e]Hawaii Facts and Figures (2009) dbedt.hawaii.goc/economic/
www.papahanaomokuakea.gov
[f]Lai BV, Fortune K (2000) The Pacific Islands, an Encyclopedia. University of Hawaii Press, Honolulu
[g]https://en.wikipedia.org/wiki/Mariana_Islands
[h]Islands.unep.ch.CHD.htm
[i] en.wikipedia/org/wiki/Samoan_Islands
[j]www.nationslonline.org/oneworld/solomon_islands.htm
[k]Tonga: Geography The World Fact Book www.cia.gov/library/publications/resources/the-world-factbook/geos/tn.html; https://en.wikipedia.org/wiki/Tonga
[l]https://en.wikipedia.org/wiki/Tuvalu

data permit. There are 260 atolls in the tropical Pacific region. In addition, there are 668 inhabited or main islands in this list, although 45% of these are in the sparsely populated Solomon group where there are only six main islands and nine atolls.

High Islands

The Pacific region is dominated by geological forces that result from the movement and interactions of its tectonic plates, relatively thin regions of the Earth's crust that lie atop a denser solid rock, the mantle, which occupies most of the Earth's interior. The principal plate in the Pacific (the Pacific Plate) underlies most of that ocean. This plate rotates and moves to the northwest, and as it does so it collides with adjacent plates on the opposite side.

There are two basic ways in which high volcanic islands form in the Pacific. The first of these occurs at colliding plate boundaries called subduction zones. Here, when one oceanic plate meets another, the portion of the denser plate is forced downward creating a deep ocean trench. The friction, pressure and water released by the descending plate causes some rock to melt and form a semifluid magma that eventually rises to form a group of volcanic islands. These characteristically form an island arc at the edge of the upper plate. Such formations are common in the Pacific and include the Mariana Islands along the border of the Pacific and Philippines Plates, and the New Hebrides (Vanuatu) arc at the same plate's northeastern and eastern boundary with the Pacific Plate. Islands can also be formed behind the edges of the subducted plate as it becomes deformed and folded into complex terraces and ridges. This is a simplified version of how the island of New Caledonia formed within the Australian Plate, making it a continental island (Fig. 1.8). Other islands have a more complex history. The high islands of Tonga and Fiji for example, constitute island arcs that appear to be associated with their own microplates that at various times have become disengaged from the edge of the Australian Plate, inducing motions separate from it [4]. Similarly, multiple small plates, each with their own rotation surround the Bismarck and Solomon Islands and the island of New Guinea at the northern end of the larger Australian Plate [5]. On many of these high islands coral reefs have been uplifted in stages exposing reef growth as it occurred tens and even hundreds of thousands of years ago.

The other primary mechanism by which high volcanic islands form is by the occurrence of hotspots, volcanic regions underlying mantle regions that are anomalously hot compared with the surrounding mantle and are thought to break through the crust. Their placement is often within the interior of a plate, but they can also occur on or near plate boundaries. Hotspots account for the chain of islands rather than arcs that often (but not always) display a sequence of movement with their associated plate over time so that the older volcanic islands are found behind the currently active volcano. The Hawaiian Islands appear to have tracked a Hawaiian hotspot that is now active on the Island of Hawaii, and has been so since 1983. The Island of Maui to the northwest is 1.3 million years old and the island of Oahu,

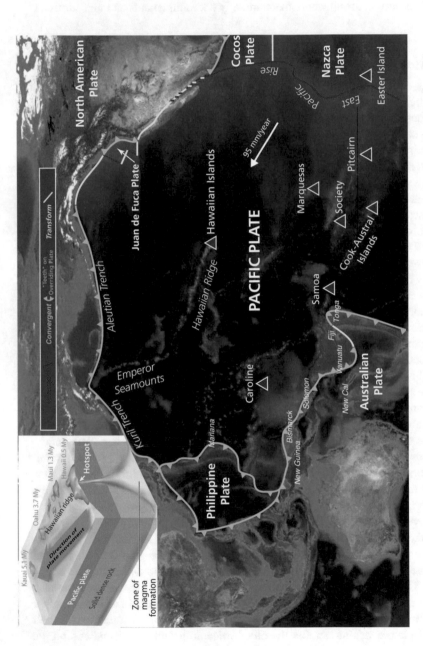

Fig. 1.8 Large plates of the Pacific Ocean are shown moving away from each other (*yellow lines*), colliding with each other (*red line*) or sliding past each other (*green lines*). Colliding plates produce trenches and island arcs along their edges, with archipelago names shown in italics. *Red triangles* depict areas of proposed hotspots. Microplates extending southeast from New Guinea are not shown. *Blue lines*, the Tropics of Cancer at the north and Capricorn at the south, depict the geographic tropics. Graphic modified from the US National Park Service. *Inset*: Hawaiian hotspot and formation/ages of the Hawaiian Islands (After NOAA Ocean explorer and US Geological Survey)

where Honolulu is located, is 3.7 million years old. The farther to the northwest one travels along the chain of islands, the older the islands become (Fig. 1.8 inset). For reasons described later, the older islands in the chain, those 30 to more than 80 million years old, have gradually sunk deep below the sea surface. These drowned and inactive volcanoes are now referred to as the Emperor Seamounts. There are perhaps 13 other suspected hotspot chains within the Pacific Plate (Fig. 1.8), but most lack sufficient data, activity or chain length to classify them with certainty. The Samoan hotspot, currently active underwater, is likely responsible for the five major Samoan Islands; likewise the Society Islands including Tahiti are associated with an intermittent hotspot but the track is quite short, especially compared with the Hawaii-Emperor chain. In some cases such as the east-west trending Caroline Islands it is not clear whether their origin can be ascribed to a hotspot or to subduction, as the volcanic islands can be associated with the movement of the small Caroline Plate east of the Philippine Plate [6, 7]. The smaller plates are not shown in Fig. 1.8.

High volcanic islands in the Pacific region can be active, dormant or extinct. Some are high enough to be capped in snow or glaciers including Puncak Jaya on the island of New Guinea (4900 m elevation), and although it is inactive there are 56 other volcanically active volcanoes there. The better-known peaks of Mauna Kea and Mauna Loa are the loftiest of the five that compose the island of Hawaii. Both are snow capped, dormant or recently extinct, and just a bit under 4300 m high. Most oceanic high islands in the Pacific are half those heights. Mt. Orohena, an inactive, weathered and dissected volcano on Tahiti at just over 2100 m elevation is an example. Mt. Waialeale on the Hawaiian island of Kauai is only about 1500 m high, but due to peculiarities of its shape and position of the moisture-laden trade winds, the rainy side of this island mountain receives 11 m of rainfall per year. This makes it one of the rainiest places on Earth. The flora of tropical high islands is dependent upon location, climate, physical diversity, island size and age. The large and ancient continental island of New Guinea has the highest diversity of plant life. The Solomon and Bismarck islands are younger, smaller and represent a relatively impoverished version of New Guinean flora, although they exhibit species that are island *endemics*, that is, found nowhere else. The Hawaiian Islands are the most isolated archipelago in the world, and because of that its flora has evolved into very unusual forms. Of the 956 flowering plants that are known 89% are endemic to that archipelago [8]. There are also 48 bird species that are unique to the Hawaiian Islands, although there were more than 100 in times past (see Chap. 8).

At the highest altitudes, tropical mountains are dry grasslands, sometimes mixed with drought-resistant shrubs and small trees. Tropical mountains at altitudes between 500 and 2000 m may be covered with clouds causing the plants to drip with condensation, especially on the side exposed to moist oceanic winds (Fig. 1.9). The vegetation is typically composed of small trees, ferns and mosses that constitute a cloud forest environment. The opposite side is often drier, and may develop a grassy broad-leaved savannah. In seasonally dry environments a dry evergreen forest may be found. Cloud forests and dry forests or grasslands are typical of the high slopes of Vanuatu, New Caledonia, Fiji and Hawaii. Lowland forests on larger islands are

Fig. 1.9 A cloud forest on the windward slope of the Hualailai volcano, Island of Hawaii (Courtesy of Riley Duren, NASA Jet Propulsion Laboratory)

found in valleys and can extend on mountain slopes to an altitude of several hundred meters depending on local conditions. They are represented by multilevel rainforests with large trees and vines forming a canopy 20–40 m above the forest floor and a rich diversity of plants below. Rainforests are well developed on New Guinea and the high islands of Fiji, but are relatively poorly developed on the smaller islands. There are a few pockets of rainforest left in the volcanic Hawaiian Islands but nearly all of them have been logged and converted to agriculture or housing developments.

Low Islands

The term 'low islands' does not convey a specific scientific meaning because there are many distinct types of low islands formed by different processes. However, the distinction of high and low islands is not unique to this book and it is useful for a comparative perspective, even if the boundaries are a bit arbitrary. If high islands are those that currently 300 m or more above sea level, low islands are considerably less than that. However, a large number of low islands are volcanic, especially those that dot Pacific Plate, and while they may have once been high above sea level, these volcanic peaks are now primarily or entirely underwater. For example, the Caroline Islands just north of the equator in the western Pacific includes a group of more than 600 islands that jut from the bottom of the Pacific Ocean. Several main islands in the group remain volcanic above sea level at altitudes of 500 m to more than 700 m and are relatively small high islands. Most of the rest are low islands capped with thick layers of limestone at or near sea level due to coral reef growth. In other places, coral reefs that grew atop volcanoes at sea level have been uplifted by plate activity, in some cases 50 m or more. These processes will be described in more detail below. In addition many Pacific islands occur on continental shelves and have been built by the accumulation of sediment as shoals, or have developed as coral reefs without the participation of volcanic predecessors.

Low islands have poor, sandy soil and little fresh water, which makes them difficult places to live. They do not support human habitation as well as high islands, and at least above water, they tended to be less interesting to the early explorers. Because of the relative uniformity of conditions, vegetation is more limited on low islands compared with the higher ones although the same types of vegetation may develop in overlapping types of habitat. However, on the low islands that developed some elevation (e.g. those in the Caroline Islands), there is little doubt from early descriptions that mature rainforests, or at least wet forests of some type had at one time become developed, but most of these no longer exist [8]. Vegetation above the high water mark typically begins with creeping plants and beach grasses followed by salt-tolerant shrubs and a few species of broadleaved trees. Mangrove communities dominated by salt-tolerant trees and shrubs develop on some islands where the coastline forms bays and is sheltered from the surf, although they are more common on high islands adjacent to river outflows (Fig. 1.10). In a few island environments with high rainfall and impeded drainage, lowland freshwater swamps may form (see Fig. 1.21d).

1.3 Coral Reefs in the Pacific

Even though they occupy between 0.2% and 0.1% of the world's ocean surface, coral reefs are likely the most biologically diverse of all habitats on planet Earth. They provide a home for perhaps 25% of all marine organisms, including almost 1/3 of all fish species. More than 100 countries have coastlines with coral reefs and tens of millions of people depend on coral reefs to at least some degree for their livelihood and their protein. Coral reefs are built by limestone-secreting animals, primarily corals and their resident algae, that are typical of clear tropical waters, but it is estimated that a third of all coral species, the architects of the reefs themselves, are now at risk of extinction from local pollution, poor land management, and from global warming (Chap. 8).

Charles Darwin was the first to comprehensively examine and classify the structure of coral reefs from his observations and those of others after his five-year voyage (1831–1836) on the relatively small HMS Beagle (Fig. 1.11). When Darwin was not seasick, and he often was, he was able to write copious notes that he turned into a treatise called The Structure and Distribution of Coral Reefs [9]. Darwin recognized three principal types of reefs: fringing, barrier, and lagoon islands (atolls). Fringing reefs are the most common of the three and are characteristically formed as narrow ribbons (fringes) close to shore (Figs. 1.12a and 1.22). Most reefs of the Mariana Islands, the Solomon Islands and the Hawaiian Islands are of the fringing type. If a 10–100 m deep trough-like depression should form parallel to shore, it becomes a lagoon that separates the shallow reef from the shoreline. That is a characteristic of a barrier reef, the second main type that Darwin described. Barrier reefs

Fig. 1.10 Low island vegetation. (**a**) *Scaveola taccada* (beach cabbage) from Maui and (**b**) *Tournefortia argentea* (tree heliotrope) from Tonga, both common shrubs or trees on Indo-Pacific beaches; (**c**) *Casuarina equisetifolia* (ironwood, Australian pine) from Efate, Vanuatu; (**d**) coconut palms (*Cocos nucifera*) are common on low islands, but these on Abaiang Atoll in the Gilbert Islands have been planted; (**e**) mangroves are common on embayed shorelines, especially on river deltas as these in Papua New Guinea (Photos courtesy of Wikipedia.org)

Fig. 1.11 HMS Beagle, a Royal Navy 27 meter-long, two-masted square rigger in the Straits of Magellan 1834 (Courtesy of Wikimedia)

are structurally more complex than fringing reefs and typically grow close to the surface where they literally form a wave barrier between the lagoon and the shoreline. They are the least common of the three reef types. Australia's Great Barrier Reef is the most referenced example although it is more than 800 km long and is not a single structure but instead is composed of about 2900 individual reefs and 300 islands. This enormous complex was discovered and charted by the British explorer and navigator extraordinaire Captain James Cook in 1770. More about Cook is described later.

There are other barrier reefs, but they are considerably smaller, and most of them surround islands rather than shadowing continental shores. The Pacific island of New Caledonia is nearly surrounded by a barrier reef more than 1500 km long, and in some places it exists in the form of double or multiple barriers. Similarly, the southeast portion of New Guinea has three barrier reefs that are complexly woven together along a distance of more than 1100 km [10] (Fig. 1.15). Other significant, but structurally simpler barrier reefs are found around Palau in the Caroline Islands, and Bora Bora in French Polynesia (Fig. 1.12b), which Pulitzer Prize-winning author James Michner thought was the most beautiful island in the world. Indeed, Bora Bora is often advertised as 'la plus belle île du monde'. There are lists of plus belles îles, for example those published by U.S. News and World Report and Condé Nast, and Bora Bora is only sometimes on them. Apparently not everyone has James Michner's taste in îles.

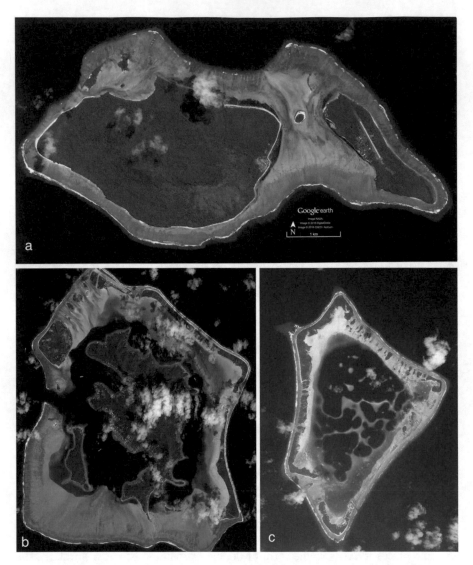

Fig. 1.12 (**a**) Yathata Islands, Fiji and their fringing reefs; note absence of lagoon. (**b**) Bora Bora in French Polynesia showing the outer barrier reef edged by white wave breaks, and peripheral low-lying barrier islands. The deep blue waters of the lagoon surround two volcanic islands, the remnants of a single extinct volcano in the lagoon center. Mount Otemanu is the highest peak at 727 m. (**c**) Atafu Atoll, Tokelau Islands, central Pacific. Atolls produce a central lagoon, but no volcanic islands protrude from it (Satellite photo (**a**) courtesy of Google Earth; (**b**) and (**c**) courtesy of the Image Science and Analysis Laboratory, NASA-Johnson Space Center)

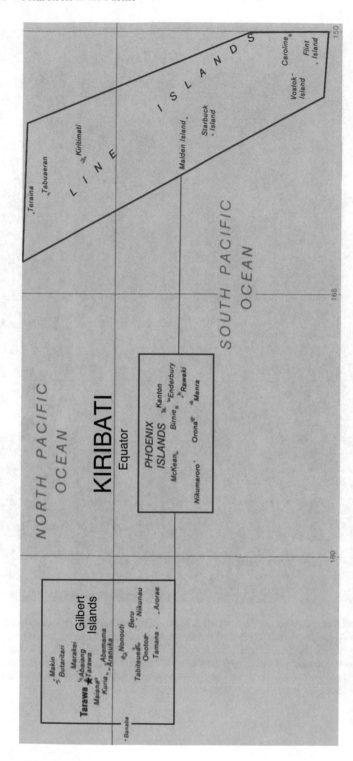

Fig. 1.22 The central Pacific island nation of Kiribati (pronounced 'Kiribass') includes three main groups of islands outlined in red: the Gilbert Islands, the Phoenix Islands and most (but not all) of the Line Islands. The latter two groups have several examples of atolls whose lagoons have become isolated from the surrounding ocean during the last 4000–6000 years due to lowering of sea level in those regions. This phenomenon underpins the process of island emergence (Map courtesy of Wikipedia.org)

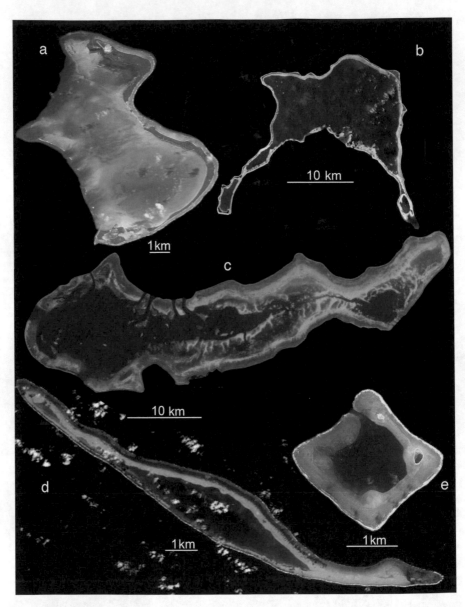

Fig. 1.13 Atoll form. (**a**) wave-shaped Onotoa Atoll, Kiribati (Gilbert Islands); (**b**) bonnet-shaped Arno Atoll, Marshall Islands; (**c**) shark-shaped Karang Kaledupa Atoll, Sulawesi, Indonesia; (**d**) spindle-shaped Barque-Canada Atoll, South China Sea; (**e**) box-like Rose Atoll, American Samoa; Note that Kaledupa and Barque-Canada are island-less atolls called atoll reefs (Satellite photos (**a**), (**d**) and (**e**) courtesy of the Image Science and Analysis Laboratory, NASA-Johnson Space Center; (**b**) and (**c**) Google Earth)

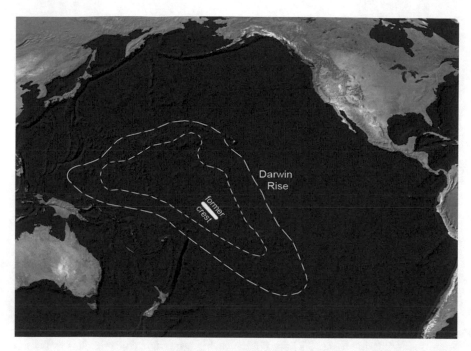

Fig. 1.14 The Darwin Rise is an area marked by volcanoes that were active 100 million years ago when the Rise was nearly 2 km above the sea floor. Its remnants include a majority of the world's atolls stretching from the Marshall Islands to Tuvalu in the northwest, Tokelau and the Cook Islands at mid-hypotenuse, the Tuamotu Islands in the southeast, and the Line Islands on the northeast. The former crest of the Rise appears to lie near Manihiki and surrounding atolls of the northern Cook Islands

Lagoon islands are the third major type of reef that Darwin described. They later became more commonly known as *atolls*, a word adapted from the language spoken in the nation of the Maldive Islands (Indian Ocean) where most of the islands occur in a ring-like pattern. Nevertheless, atolls are even more abundant in the tropical Pacific. The rings are formed in part by islands that more or less enclose a central lagoon, although "ring-like" does not necessarily mean circular, and in fact, atolls generally tend to be more similar to ellipses [11]. However, a survey will reveal a nearly infinite variety of shapes. Atafu Atoll in the Tokelau Islands (Fig. 1.12c) for example is polygonal. South Minerva

(Tonga) is a figure eight. Onotoa in the Gilbert Islands is wave-shaped (Fig. 1.13a). There are pear- and ham-shaped atolls, heron-shapes, and those that look like a bonnet or a spindle (Figs. 1.13b and d). There are even some that resemble a legless goose or a shark (Fig. 1.13c). Rose Atoll (Fig. 1.13e) is square. A summary of atoll locations, shapes and other information can be found on the author's website [12].

Charles Darwin and the Evolution of Atolls

More than 80% of the world's atolls are found in the Pacific and adjacent seas (Figs. 1.14 and 1.15), and most of those are found in an area of the central Pacific called the Darwin Rise. Here extinct volcanoes tower nearly 2 km above the ocean floor where they sprout from an elevated blister-like triangle of oceanic crust thousands of kilometers across. This remote area is not only atoll-rich, but is home to thousands of other islands that produce fringing and barrier reefs.

Although they have become pushed up by volcanic activity beneath them, atolls do not show any trace of igneous rock. All of that volcanic material has been covered with reef limestone (carbonate) produced by animals and plants. Young carbonates in turn, have accumulated in distinct layers on top of older limestone material, indicating that reefs have been growing atop their volcanic platforms for a very long time. Darwin did not have the specific geophysical evidence, but he hypothesized that the volcanic mountains were undergoing a type of gradual sinking process called *subsidence,* the basis of which is the gradual cooling and plate movement of the oceanic crust on which the volcanoes formed. Darwin knew that islands could either be lifted up or sink beneath the surface, but he knew nothing about mobile tectonic plates. And he also did not know about the effects of past global climate and sea-level change.

Darwin reasoned that as subsidence continued, vertical coral growth should keep pace with the rate of sinking. Thus, coral reef limestone on volcanic platforms should be rather thick, not merely a thin veneer over a volcano as was widely assumed. He suspected from coral growth on sunken ships that corals could grow more quickly than many thought, but he did not know either the rates of sinking or of coral growth. Mr. Darwin needed data he did not have, but that makes his observational powers all the more prescient. In a similar and equally remarkable manner, Darwin predicted the process of evolution without knowing anything about genes or genetics, molecules or molecular biology, or the many other branches of science that underpin the process. Eventually both hypotheses would move on to being generally correct as originally stated, but they were modified and supported by subsequent work- the mark of graduation to a scientific theory. This is the same Darwin who studied geology and natural history at Cambridge University, but whose Bachelor of Arts degree was intended to prepare him for a career as a church parson after his voyage on the Beagle.

The subsidence hypothesis was elegant in its simplicity: Coral reef growth is restricted to shallow, tropical waters, mostly less than 50 m deep. But if reef

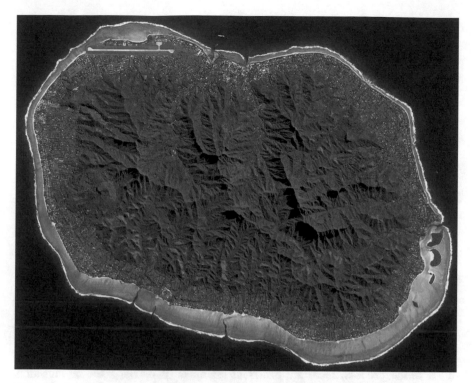

Fig. 1.16 Rarotonga in the Cook Islands is volcanic and extends nearly 4600 m from the Pacific sea floor and is 195 m above water. It is surrounded by a fringing reef notwithstanding a small lagoon and barrier islands at the southeast side. The predominant winds are from the east and southeast, making that the windward side of the island. The surf marks the reef crest and the reef flat is shoreward of that area. The runway at the island's northwest is 2246 m long (Photo courtesy of NASA and Wikipedia.org)

limestone can be deposited fast enough to keep pace with subsidence, corals could remain in shallow water. Therefore if the volcanic platform has been subsiding at fractions of a cm per year for millions of years, coral reefs should be thousands of meters thick, and the volcanic basement, once at the surface, should now also be thousands of meters below the limestone. As it turns out, the subsidence rate of the Pacific plate is nearly 30 cm per thousand years, easily within the range of coral growth in most parts of the tropics, including the Pacific where the average reef growth over the last 5–10 thousand years has been half a meter or more per thousand years [14].

Darwin then went a step further and asserted that fringing reefs, barrier reefs and atolls are all related in a developmental sequence. As corals initially colonized a volcanic peak, a fringing reef would form as shown surrounding the island of Rarotonga in the Cook Islands (Fig. 1.16). Further subsidence and upward coral growth would account for a lagoon around the remaining volcanic peak. Eventually the volcanic peak should disappear entirely beneath the surface and an empty atoll

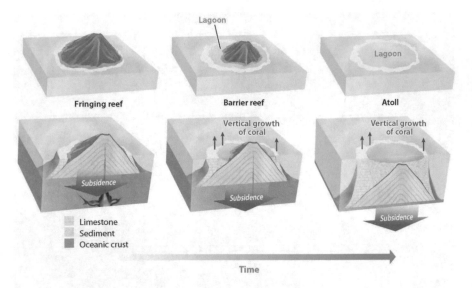

Fig. 1.17 Darwin's hypothesis of atoll formation: A fringing reef growing on a subsiding volcanic platform gives rise to a barrier reef and lagoon followed by atoll formation when the volcanic material disappears (Figure from Goldberg 2013 [15])

lagoon surrounded by a ring-like reef would remain (Fig. 1.17). As a further testament to the process, there are excellent examples of barrier reefs on subsiding islands that have not quite achieved the atoll stage. In such cases the volcanic peak or peaks remain protruding in the lagoon. Quite appropriately, the name of 'almost atoll' is given to this stage. Examples are shown in Fig. 1.18. Of course 'almost' places time in a geological context. On Maupiti (western French Polynesia), the volcanic peak is nearly 250 m above sea level in the lagoon, and assuming a constant and maximal rate of subsidence, it should take somewhere close to 800,000 years for it to disappear. At nearly 90 m we might expect Exploring Isles' peaks (Fiji) to take longer. The same would be expected of Chuuk (Caroline Islands) and Mangareva (eastern French Polynesia), which are about 130 m above sea level. Aitutaki at 35 m (Cook Islands) should take a bit less time to disappear than the others.

Fourteen years after Darwin's death in 1882 the British Royal Society mounted an expedition to Funafuti Atoll in what is now the nation of Tuvalu (formerly the Ellice Islands). After a few failures they managed to drill to a depth of more than 360 m, and the cores they recovered contained carbonates including coral material that could only have grown in shallow water. Darwin appeared to have been right, but no volcanic material was in the cores. So how deep did the limestone subside, and where was the surface of the volcano? The answer to that question was beyond nineteenth century technology. However, in 1951–52 the United States began conducting tests on the geology of the Marshall Islands as part of a military program of atomic weapons research, and several boreholes were drilled. The deepest of these eventually penetrated shallow-water limestone on the atoll of Enewetak to depths of

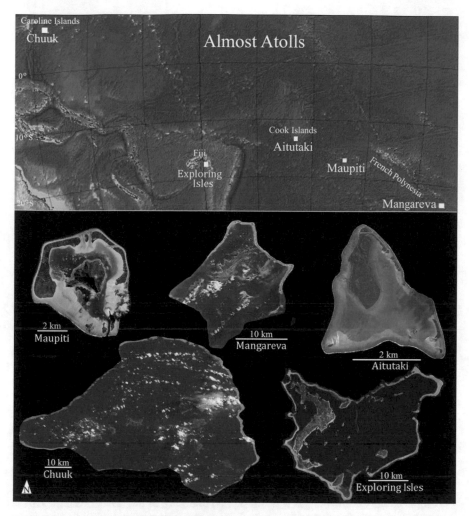

Fig. 1.18 Almost atolls display their volcanic peaks within the lagoon. With continued subsidence and coral growth they will become atolls. Satellite views from Landsat, U.S. Geological Survey

1250 and 1400 m (Fig. 1.19). It was at those depths that the volcanic base was found. The cores from the Marshall Islands contained fossil corals suggesting that reefs had been growing discontinuously for millions of years. We will get to the 'discontinuous' part a little later.

Even though Darwin was right about subsidence, it is probably not just subsidence and coral growth that is responsible for the shape of atolls, especially the ones with irregular configurations. Instead, various authors have suspected that there are additional, if not completely alternative explanations for the shape of atolls and lagoons. Several investigators have postulated that exposure to rainfall during glacially induced low stands of sea level is a means of wearing away limestone [16].

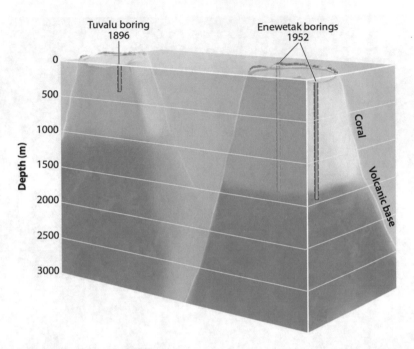

Fig. 1.19 The British Royal Society was the first attempt to drill through an atoll in Tuvalu in 1896 and the cores that were brought up from more than 360 m were all filled with shallow-water carbonates; there was no volcanic material. In 1952 the United States drilled through shallow-water limestone to more than 1250 m before finding the volcanic surface, thus Darwin's subsidence hypothesis was vindicated 120 years after he proposed it (After Goldberg 2013 [15])

Rainfall will etch and dissolve carbonate through its relative acidity (compared with that of sea water), and exposure to it for tens of thousands of years at a time may be consequential for the stability of the limestone. Other forces such as tsunamis or earthquakes may contribute to weakening and cracking of the limestone cap. This is suspected of causing a failure in the reef structure, perhaps causing a section or sections to fall away. Such landslides on the sharply sloped sides have been observed on some atolls and may initiate restructuring of the atoll margin, thereby changing its original oval or elliptical shape [17]. The missing sections of Johnston (central Pacific) and Mururoa (French Polynesia) atolls are almost certainly due to the production of faults, and the limestone debris on the flanks of those atolls attests to alteration of the reef by collapse of those parts of the ring. It may not be entirely coincidental that these atolls were among those used for nuclear weapons testing in the twentieth century (Chap. 7).

Atoll Lagoons

The lagoon constitutes a far greater proportion of most atolls than the land area. Boomerang-shaped Kwajalein in the Marshall Islands is the largest atoll to be found on the Darwin Rise. At more than 8 km across, covering an area of about 2600 km^2, the lagoon is about 60 m deep. The total land area is about 15 km^2. In general, there is a positive correlation between atoll size and lagoon depth, but there are exceptions such as Nukuoro in the Caroline Islands. This small atoll is about 6 km across, but its lagoon is nearly a bottomless pit, more than 100 m deep [16]. Nukuoro should not be confused with Nukunonu in the Tokelau Islands, even though it too is a small atoll. But while it is twice the size of Nukuoro, its lagoon is far shallower. Perhaps this is good time to mention that Polynesian names can be confusing. Nukufetau and Nukulaelae are in the Tuvalu Islands, whereas Nukutoa is in the Bismarck Archipelago; Nukutipipi is in French Polynesia and Nukumanu is an atoll in the Solomon Islands. Nukumbasanga and Nukusemanu are in Fiji. And those are just the atolls that start with 'nuku', a Polynesian prefix for 'island'. Nukumanu, for example, means 'island of birds'; Nukutoa means island of courage (or war or defiance).

As Darwin wrote in his diary aboard the Beagle "If we imagine such an island, after long successive intervals to subside a few feet the coral would be continued upwards. In time the central land would sink beneath the level of the sea and disappear, but the coral would have completed its circular wall. Should we not then have a lagoon island?" Darwin thought that reef corals on the rim simply grew faster than those of the lagoon and more closely tracked subsidence rates. In fact, Darwin knew nothing about large changes in sea level (more than 100 m) due to withdrawal of water during glacial periods, and its reversals afterward, and he could only guess at the time scale for atoll formation (tens of millions of years). Changes in sea level were responsible for the discontinuous growth of atolls noted during the coring process on Funafuti and Enewetak. But what if changes in sea level better account for lagoon formation, not by just coral growth and subsidence as Darwin suspected, but as the result of exposure during glacial periods? The tops, now exposed to the atmosphere are likely to accumulate rainfall, especially near the center, and begin to dissolve limestone there. In fact, there is a good relationship between latitude and rainfall on one hand, and the area and maximum lagoon depth on the other, the average of which is a little over 30 m [16]. Lagoon depths may be better accounted for as solution features than by coral growth, even though the latter classical view is often taken as an article of faith for which there is little direct evidence. There is also a third compromise theory, evidence that the diversity of form we see today results from a combination of factors including coral growth, subsidence, and sea-level changes acting together [17, 18]. Lagoons are dynamic systems that may fill with sediment and become shallower with time or can be flushed and cleaned (see below). Few lagoons have been adequately investigated to enable choosing among the models for their development.

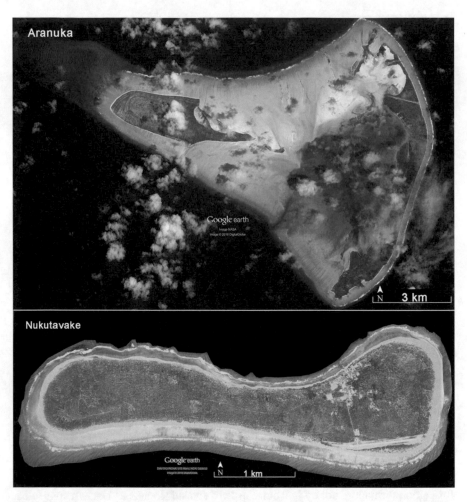

Fig. 1.20 A lagoon with a sand apron up 1800 m wide around the island at the west and a diminished, relatively shallow lagoon (about 18 m deep) on the east characterizes Aranuka Atoll (Gilbert Islands, central Pacific) and suggests active sand accumulation. Nukutavake is a former atoll whose lagoon has been filled by sediment deposition and has disappeared (Images courtesy of DigitalGlobe)

Large and relatively deep lagoons such as those in the Maldive Islands (Indian Ocean) may be flushed by currents and swept clean, even though some sediment may be transported into the lagoon. In such cases water and sediment export to the open sea is often referred to as a leaky bucket atoll, a metaphor for a lagoon that will not likely fill. However the presence of a sand apron at the outermost edges of the lagoon is a ubiquitous feature of atolls (see Fig. 1.20) and is indicative of some degree of sediment transport to the lagoon. Indeed, small atolls with shallow lagoons are especially prone to becoming gradually filled by incoming sediment. That

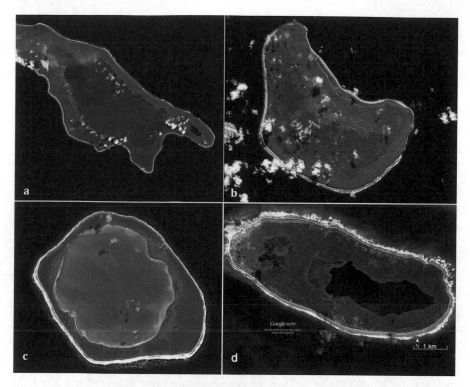

Fig. 1.21 Uplifted and emerged atolls. (**a**) the eastern end of Rennell Island, part of the Pacific's Solomon Islands group, contains a brackish-water lagoon, but it has been uplifted to more than 180 m above sea level. (**b**) Makatea is a former atoll in French Polynesia that has been uplifted more than 100 m by nearby volcanic activity. (**c**) Niau, also in French Polynesia, is elevated to about 6 m but despite being physically closed from the ocean its lagoon water is gradually exchanged through fissures and crevices in the surrounding limestone. (**d**) Teraina Island (Line Islands, Republic of Kiribati) has been exposed by lowered sea levels since its formation. Now 3–5 m above present sea level, its lagoon is fresh water and is surrounded by bogs and marshes (Satellite views (**a**) courtesy of Landsat; (**b**) from NASA, (**c**) from the European Space Agency and (**d**) from DigitalGlobe)

dynamic depends on a number of features including storm frequency, island distribution and the height of the tides. Aranuka Atoll in the Gilbert group (Fig. 1.20), for example, has a very broad and well-developed apron that extends from being exposed at low tide, to a depth of 2 m where sediments are transported by high tides, and the lagoon is diminished accordingly. Nukutavake in the Tuamotu group is a former atoll whose lagoon is completely filled and is now occupied by vegetation (Fig. 1.20). Several former atolls in French Polynesia, Tuvalu, and others in the Indian Ocean have been converted to lagoonless islands in this manner.

Uplifted and Emergent Atolls and Islands

Even though its cliffs rise more than 180 m above sea level, Rennell Island (Fig. 1.21a) still maintains a lagoon at one end of the island, although it is now a brackish-water lake. Rennell's uplift appears to have been caused by a bulge associated with the formation of the nearby New Hebrides Trench, a huge chasm more than 7600 m deep caused by the collision and over-riding of the Australian Plate by the Pacific Plate. Uplifted atolls are usually not elevated to this extent. Makatea in French Polynesia, for example (Fig. 1.21b) is raised by only half as much as Rennell, but as is also typical in such cases, the lagoon's contact with the ocean has been lost and the depression is now filled by terrestrial vegetation. This former atoll was uplifted by plate flexure associated with the now extinct volcanic activity of the nearby Society Islands, a group that includes Tahiti. Niau (Fig. 1.21c) is another atoll in French Polynesia that was elevated to a more modest 6 m above sea level by the same flexure that affected Makatea, but Niau is farther from the plate boundary. Some atolls and islands have been lifted out of the water due to movement of the seafloor and as the result of erosion, high cliffs often surround their outer edges. Uplift occurs due to depression of a tectonic plate by the formation of a volcano, which in turn causes the plate to flex and bend nearby. Alternatively, plates in collision can cause uplift as the one plate dives under the other. Flexure causes a bulge in the over-riding plate that can force atolls above sea level. An example of the latter case is Rennell Island, known better for its place as the last major battle at sea between the United States and the Imperial Japanese Navy during World War II, is a most spectacularly uplifted atoll. The high walls of the atoll prevent direct contact with the surrounding ocean. The lagoon is therefore uplifted and all but closed to seawater exchange, although an indirect and lengthy renewal process occurs that depends on storm overwash and leakage of seawater through the porous the reef limestone. As a consequence the color of the water is distinctively green due to the abundance of algae and green bacterial mats, which do quite well in this environment, but the biological diversity of marine animals is quite low [6, 15].

A number of other atolls in the central Pacific region, especially those within the island nation of Kiribati (Fig. 1.22), have found themselves out of water. Teraina (Fig. 1.21d) and several other atolls in the central Pacific have also become exposed, but not due to uplift. Some atolls that flourished several thousand years ago were built to a sea surface that was a meter or more higher than it is now due to the intrusion of glacial melt-water, a process that continues to elevate sea level today, although for now at a slower pace. During the last 4000–6000 years, sea levels have remained the same in some areas and re-adjusted in others. In the central Pacific, an 'equatorial siphoning effect' occurred in which glacial melt-water was gradually drawn away to fill the deformation of the seafloor caused by the former load of ice at the poles. Thus, sea levels in that region fell 1–2 m [19]. An additional factor contributing to this effect is subsidence, which in some areas has either become irregular or has stopped entirely. Because of this, atolls such as Teraina and several others in the central Pacific have become *emergent atolls*, those that grew when sea

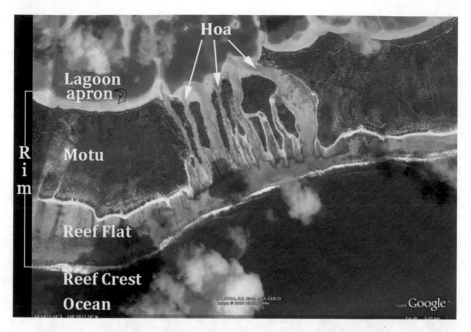

Fig. 1.24 The reef rim extends from the reef crest to the lagoon apron and includes its reef flats and islands. Shallow channels that traverse the rim between islands (motu) are called hoa in Polynesian, as shown on the atoll of Mataiva, French Polynesia. Most of these hoa are open, but asterisks mark two that are closed at the lagoon end (Image courtesy of DigitalGlobe)

levels were 1–2 m higher than they are now, and have been left to protrude above current sea level.

Coral Reef Zones

Despite considerable variability when different types of coral reefs and reefs from different places are compared, there is a generalized structure that can be observed. Coral reef growth is limited to relatively shallow water. Under clear oceanic conditions that depth can be 100 m or more, although it is more often less than that. This deep-water zone is called the *outer reef slope* and on oceanic islands the angle of descent can be extremely sharp, but less so as this zone reaches shallower water. Here several other zones become evident, including (from the ocean to the lagoon) the reef crest, reef flat, the islands, and the sand apron of the lagoon (described above). Sometimes these shallow structures are lumped together and referred to as the *reef rim* (see Fig. 1.24). The *reef crest* is the reef's uppermost point of growth and is limited by sea level. This zone is at least partly uncovered at low tide, revealing its composition of heavily calcified red algae (Fig. 1.23). These particular forms produce a nearly indestructible reddish pink *algal ridge* composed of dense

Fig. 1.23 The pink surfaces of calcified red algae are exposed at low tide and form the reef crest on the windward side of Rarotonga, Cook Islands. The crest extends in places on the surfaces of the reef flat along with a few hardy corals. Note the incoming heavy surf in the background and tide pools formed in the reef rock (Photo courtesy of Genny Anderson, Santa Barbara City College)

carbonate sheets that cement this section of the reef together. Although a bit coun-terintuitive, the reef crest and its algal ridge are typically enhanced on the *wind-ward*, more energetic side of the island, making them more prominent there. This is because water motion replenishes food and nutrients for reef animals and plants. It also removes sediment and moderates extreme temperatures. A few hardy types of coral also exist on or near certain reef crests by forming rounded, flat, or other hydrodynamic shapes that decrease their profile to incoming waves. Conversely, a reef crest and other zones may be absent on island shores that face away from the wind, the *leeward side* of the island.

The *reef flat* is the next of the rim zones and it lies behind the crest closest to shore. Flats are only slightly sloped as the name implies. This zone is very shallow and can be uncovered at low tide, but the living things that may be found on them can vary significantly according to the width, topography and slope of the shore, as well as with tidal height and wave energy it is exposed to. On the windward side the reef flat is often rocky due to natural topography or coarse reef rubble generated by storms. The calcareous algae of the reef crest can also extend into the reef flat, forming cracks, crevices, pools and troughs that hold water and shelter a variety of plants and animals (Fig. 1.23). Without such refuges they would otherwise be exposed to high light, high temperature and diluting rainfall. By contrast, medium and fine coral sand may accumulate in calmer environments such as the leeward side or other protected island shores.

The Reef Rim and Coral Islands

Islands may form on barrier reef and atoll reef flats and are typically only 3–4 m above sea level, although windblown dunes covered with vegetation can reach 6–8 m. Their construction is facilitated by pre-existing rocky platforms such as beach rock that forms along shorelines, or high spots composed of wind-driven coral sand. These islands are composed of limestone (coral fragments and sand), and are not volcanic. Two types of coral islands are typical of the atoll environment and are thought to be the product of energy exposure [20, 21]. The first is called a *sand cay*. These are unvegetated accumulations of sand built under conditions of relatively calm seas. Rose atoll in American Samoa (Fig. 1.13e) has a typical small sand cay perched on its northeast corner, and another vegetated island nearby. Tiny though its islands may be, they are home to large colonies of federally protected migratory seabirds including terns, boobies, frigatebirds, and tropicbirds. Rose Atoll is a United States National Wildlife Refuge. Sand cays are prone to build up, diminish or even disappear and reappear depending on storm and swell conditions. On the other hand, vegetated islands are typically perched on reef gravel and rubble and tend to be better developed on the windward side of atolls. Onotoa Atoll's islands (Fig. 1.13a) are concentrated on the northeast side due to winds and currents that drive reef sediment toward high spots or rocky platforms. Here, sand or broken pieces of coral pile up and begin the island-building process. Animals including corals and certain types of seaweeds produce those carbonate sediments and are indicative of a healthy reef. Should the sediment supply diminish due to pollution or other insults, island formation or maintenance could be jeopardized and begin to take on the appearance of the leeward side of Onotoa Atoll. That side often lacks large-scale reef development and island formation. Most Pacific islands have developed this sort of windward-leeward asymmetry, with islands on less than half of the reef circumference. The barrier reef islands on Bora Bora for example (Fig. 1.12b)

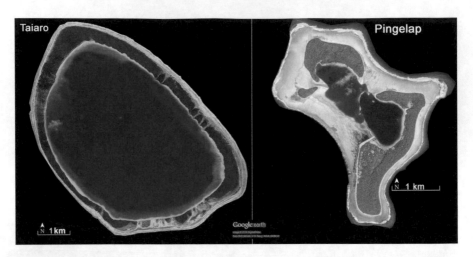

Fig. 1.25 Closed lagoons. Taiaro is a small atoll in the western Tuamotu Archipelago (French Polynesia). The rim is almost completely occupied by islands, but there are about 12 hoa. Pingelap is a small atoll in the Caroline group where islands constitute about 25 % of its rim. A wide, shallow reef flat separates the two large islands at the northeast and at the southwest, connecting the ocean with the lagoon. When the reef rim or its islands completely surrounds the lagoon it is said to be a closed atoll (Satellite views from DigitalGlobe)

present a solid wall along the windward east and northeast side but are poorly developed on the leeward south and west. However, there are important exceptions to this rule. Some atolls exposed to monsoon winds that change direction seasonally, develop islands on two or more sides. Islands surround many of the Maldive and the nearby Lakshadweep atolls off the southwest coast of India, as well as in the South China Sea for this reason.

Atoll and barrier reef islands with vegetation are often referred to as motu, their Polynesian name (Fig. 1.24), and in many cases there are gaps between islands that facilitate exchange and renewal of water in the lagoon. We could call them 'channels' but that would overlook the nuances. If the people of Lapland have 180 words for snow and ice, there should be at least a few words to describe gaps or breaks among motu. The most often used Polynesian term is *hoa*, which refers to narrow and shallow (less than a meter deep) trans-rim passages that allow ocean water into the lagoon during onshore winds or at high tides [22]. The degree of efficiency of a hoa depends on how many there are, their position, and their length as well as their width. At Raroia Atoll (Tuamotu Archipelago, French Polynesia) there are 260 of them. On Mataiva Atoll in the Tuamotu group there are 'open hoa' that extend all the way across the rim (Fig. 1.24). But there are also closed hoa that have been clogged by storm activity on either the lagoon or the ocean side, and when they are completely blocked with sediment, they become 'dry hoa'. There are even hoa that can be distinguished in old reefs called paleohoa. And there are some atolls that do not have hoa at all, which could make them hoa nele (Polynesian: lacking hoa), but there is no such term that I am aware of. Atolls with hoa that allow water into the lagoon on a regular basis are referred to as atolls with *open lagoons*.

In contrast to Mataiva Atoll, nearby Tairao (Fig. 1.25) is an atoll with a *closed lagoon*, suggesting that there is at least some restriction on the exchange of lagoon water with the open ocean. Some atolls with hoa only become functional during storms, and in others, especially those with very large and deep lagoons, have a long *renewal time*, the time it takes for all of the lagoon water to be exchanged. This can be a matter of months or even years, especially if storm frequency is uncommon. The actual time that it takes for lagoon waters to be replaced is a complex amalgamation of rim and lagoon structure, wave height, swell direction and tidal amplitude. Each atoll has its circulation idiosyncrasies that can make predictions of renewal time challenging.

Closed atolls are rather common. In fact, 78 atolls with lagoons are found in French Polynesia, and most (53) of them are closed by the definition of having a navigable passage. There are 104 closed atolls known of the 439 that have been documented. However, the definition of 'navigable' could use some tweaking. What kind of vessel or draft must there be, and how deep does the passage have to be? For example, there are some fairly narrow, 4–5 m-deep passages through the reef on Lae, Ujelang and Utrick atolls in the Marshall Islands that may make them marginally navigable [13].

Some atolls have large breaks between islands. On Pingelap (Fig. 1.25) there is a wide, rubble-strewn reef flat on the windward (northeast) side as well as on the opposite side between the two main islands. The rim of this small atoll occupies about 700 hectares, but the two relatively large islands compose only 25% of that area. The largest is about 5 m above sea level. Terrestrial Pingelap is fairly lush for an atoll, but the forests are dominated by coconuts, breadfruit and screwpine (see Chap. 4), most of which have been planted. And because it lies in the equatorial rain belt and gets an average about 4 m of rainfall per year, Pingelap is one of the wettest atolls in the Pacific.

The other 75% of the rim is composed of a reef flat unobstructed by islands, and even at low tide, there may be sufficient water to navigate into the lagoon using small boats. One suspects that a few weeks of tidal exchange may be sufficient to refresh the lagoon volume on this atoll. Even so, Pingelap is considered closed because the rim is without gaps through the reef flat. In order to qualify as an open atoll, there has to be a navigable gap through the rim of the reef, and Pingelap does not have one.

Pingelap is an interesting place for another reason. The population here is small, only about 800 people, but up to 10% of them are colorblind and another 30% carry the gene for it. This is not the garden variety of green-red color blindness; this is achromatopsia- complete loss of color vision that is often accompanied by light sensitivity, a loss of visual acuity and an uncontrollable motion of the eyes themselves. During the day many islanders lower their heads, blink and squint constantly if they go outside, and they often live with colored film over their windows. The Pingelapese have adjusted in part by shifting their outdoor activity patterns to coincide with the evening hours. They have become nocturnal. Achromatopsia is an

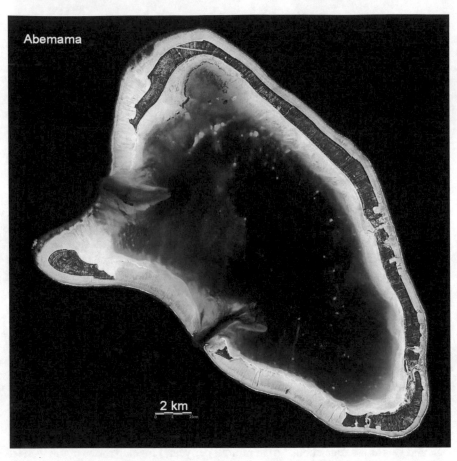

Fig. 1.26 Abemama Atoll in the Gilbert Islands has two navigable passages through the rim as shown in the SW and NW sections (Image courtesy of Landsat, U.S. Geological Survey)

inherited condition that results from defects associated with four genes, two of which are contributed by each parent. The parents with only one pair each will be carriers, but their offspring will have a one in four chance of inheriting the disease. In the United States one in 33,000 people is affected. So how did it get to one in ten on Pingelap? In 1775 Typhoon Lengkieki struck the island, a storm so fierce that it left only 20 people alive out of about a thousand. One of the survivors was the King of Pingelap and he was an achromatopsia carrier who interbred with cousins, some of whom also carried the disease genes. In fact, all achromats can trace their ancestry to the king. More importantly, this firmly inaugurated total color blindness in this small island population. On Pingelap, the 'founder effect' as it is called, is a chance event that can alter the gene pool of a population, making a rare condition common. This founder effect was made famous by neurologist Oliver Sacks in his 1998 book Island of the Colorblind [23].

Most atolls that are considered 'open' have another type of channel where there is a complete and unambiguous break in the rim. Such a passage is called *ava* in Polynesian and one example is shown in Fig. 1.26 on Abemama Atoll in the Gilbert group. Atoll *ava* are quite common. The one at the north side of Rose Atoll (Fig. 1.13e) is more than 36 m wide and up to 14 m deep, and likely has a significant effect on the amount of time it takes to replace its lagoon water.

A number of atolls only develop very small islands on their perimeter. On Rose Atoll in American Samoa, for example, there are only two islands totaling about 8 hectares, and together they constitute about 5% of the rim. On Passau Keah Atoll in the South China Sea there is only one island, a sand cay that constitutes less than 0.5% of the rim. Are there atolls that do not develop islands at all? Yes, lots of them. By my count [13], 39% of all atolls (171 of 439) do not have dry land (or very little of it), but still produce a clearly defined lagoon and a reef rim a few meters below the surface. These are called *atoll reefs* to distinguish them from atolls with islands. In some areas such as the South China Sea, and Fiji, almost all the atolls are atoll reefs, as are many in Indonesia (see Barque-Canada Reef and Karang Kaledupa, Fig. 1.13c, d). Atolls without appreciable island development are also common in the Caroline Islands and those east of Papua New Guinea.

References

1. CSIRO, Australian Bureau of Meteorology and SPREP (2015) Climate in the Pacific: a regional summary of new science and management tools, Pacific-Australia Climate Change Science and Adaptation Planning Program Summary Report. Commonwealth Scientific and Industrial Research Organisation, Melbourne
2. Chen B, Lin X, Bacmeister JT (2008) Frequency distribution of daily ITCZ patterns over the western-central Pacific. J Clim 21:4207–4222
3a. Cai WS, Santoso A et al (2014) Frequency of extreme El Niños to double due to greenhouse warming. Nat Clim Chang 4:111–116
3b. El Niño 2014–2015 (2015) Impact outlook and policy implications for Pacific islands. United Nations Economic and Social Commission for Asia and the Pacific. http://www.mfed.gov.ki/sites/default/files/UNESCAP%20Report%20El-Nino%20Potential%20Impacts%20in%20Pacific%20Island%20Countries.pdf
4. Taylor GK, Gascoyne J, Colley (2000) HRapid rotation of Fiji: paleomagnetic evidence and tectonic implications. J Geophys Res 105:5771–5787
5. Baldwin SL, Fitzgerald PG, Webb LE (2012) Tectonics of the New Guinea region. Ann Rev Earth Planet Sci 40:495–520
6. Neall VE, Trewick SA (2008) The age and origin of the Pacific islands: a geological overview. Phil Trans Roy Soc B 363:3293–3308
7. Ur RH, Hideo N, Kei K (2013) Geological origin of the volcanic islands of the Caroline group in the Federated Sates of Micronesia, Western Pacific. South Pac Stud 33:101–118
8. Mueller-Dombois D, Fosberg RF (1998) Vegetation of the Tropical Pacific Islands. Springer, New York
9. Darwin C (1842) The structure and distribution of coral reefs. Reprinted by University of California Press, Berkeley
10. Spalding MD, Ravillious C, Green EP (2001) World Atlas of coral reefs. University of California Press, Berkeley

11. Stoddart DR (1965) The shape of atolls. Mar Geol 3:369–383
12. Goldberg W (2016) Atolls of the world. http://www.maps.fiu.edu/gis/goldberg/atolls
13. Goldberg W (2016) Atolls of the world: revisiting the checklist. Atoll Res Bull 610. http://opensi.si.edu/index.php/smithsonian/catalog/book/132
14. Dullo W-C (2005) Coral growth and reef growth: a review. Facies 51:33–48
15. Goldberg W (2013) The biology of reefs and reef organisms. University of Chicago Press, Chicago
16. Purdy EG, Winterer EL (2001) Origin of atoll lagoons. Geol Soc Am Bull 21:837–854
17. Terry JP, Goff J (2013) One hundred thirty years since Darwin: 'Reshaping' the theory of atoll formation. The Holocene 23:615–619
18. Toomey M, Ashton AD, Taylor J (2013) Profiles of ocean island coral controlled by sea-level history and carbonate accumulation rates. Geology 41:731–734
19. Grossman EE, Fletcher CH, Richmond BM (1998) The Holocene sea-level highstand in the Equatorial Pacific: analysis of the insular paleosea-level database. Coral Reefs 17:309–327
20. Stoddart DA, Steers JA (1977) The nature and origin of coral reef islands. In: Jones OA, Endean R (eds) The biology and geology of coral reefs, vol 4. Academic Press, New York, pp 59–105
21. Yamano H, Kayanne H, Chikamori M (2005) An overview of the nature and dynamics of reef islands. Glob Environ Res 9:9–20
22. Stoddard DR, Fosberg FR (1994) The hoa of Hull Atoll and the problem of hoa. Atoll Res Bull 394:1–26
23. Sacks O (1997) Island of the colorblind. Alfred A. Knopf, New York

Chapter 2
Populating the Pacific

Abstract The original migrants to the islands of the western Pacific likely took place about 25,000 years ago during the last glacial period when sea levels were considerably lower than they are today. New Guinea and its offshore islands would be colonized first, and this is where the development of the Papuan language took place. However, its mountainous interior isolated and fragmented the language into multiple families including perhaps 900 distinct and mutually unintelligible language isolates, some of which are spoken only in a single village of 25–50 individuals. Archaeological and linguistic evidence suggests that the first non-Papuan migrants originating from Taiwan about 5000 years ago were spreading out in new waves of migration to coastal New Guinea and the nearby islands. These people are the likely source of a family of some 400 languages known as Austronesian. Beginning about 3500 years ago some of these Austronesian speakers began to move into islands as far east as Tonga and Samoa where they developed stilt houses along with quadrangular stone adzes, distinctive red pottery and domestic pigs. These were the Lapita people. As they and their progeny began to explore even farther, a considerable sophistication of naval architecture began to emerge along with navigational and seamanship skills. Various forms of the Austronesian language expanded as well, including as far east as Hawaii and Easter Island, and south to New Zealand. The indelicate impacts of these societies on the land are described, as are their divisions into three general cultural groups including Melanesia, Micronesia and Polynesia.

2.1 The Original Migrants and Wallacea

During the Pleistocene, a period of time from about 2.6 million to about 12,000 years ago, four major cycles of warm and cold glacial periods occurred along with corresponding sea-level changes. As glaciers enlarged at the poles, sea level dropped by at least 100 m as water was withdrawn from the ocean. Thus for periods of tens of thousands to hundreds of thousands of years at a time, sea level was sufficiently lowered to allow animal migration between the Indo-Malay mainland and the large islands of Indonesia. Likewise, a land bridge formed between the island of New Guinea and the continent of Australia. But the two connected landmasses were

© Springer International Publishing AG 2018
W.M. Goldberg, *The Geography, Nature and History of the Tropical Pacific and its Islands*, World Regional Geography Book Series,
https://doi.org/10.1007/978-3-319-69532-7_2

surrounded by impenetrably deep waters, and despite the islands remaining between them as stepping-stones, a very effective barrier was formed for land animals (Fig. 2.1). Large mammals such as orangutans, tigers and Asian elephants were able to migrate freely to the south from Asia, but could only get as far Borneo and Java. Likewise, marsupials including kangaroos and wallabies, and flightless birds such as emus and cassowaries that originated in Australia and New Guinea were unable to cross into Asia. The famous naturalist Alfred Wallace made these observations in the 1850s and spent eight full years studying the geographic distribution of animals in that region. He was also interested in how the environment seemed to influence the adaptations of such animals, and thus he became an evolutionary biologist as well as a biogeographer. Indeed, his ideas on evolution were very similar to Darwin's, and letters that the two exchanged suggested to Mr. Darwin that perhaps he ought to finally publish his Origin of Species, a manuscript he had been working

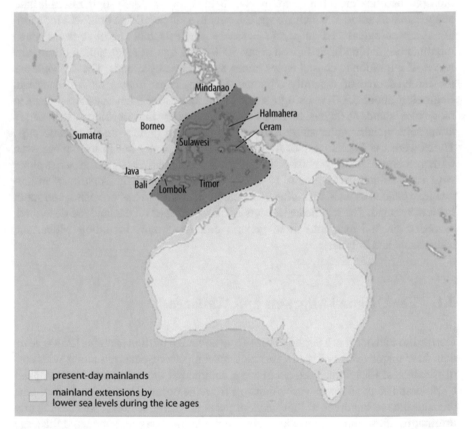

Fig. 2.1 Wallacea in red is the southern limit of typical Asian mammals, and the northern limit of Australian marsupials and other characteristic animals. The islands between are separated by the deep waters of Indonesia and represent a barrier to migration; the exposed land connecting Australia and islands including New Guinea 25,000 years ago (*colored gray*) facilitated migration (Modified from Wikipedia.Org)

on for 20 years. In the long run, Darwin became famous and Wallace became better known for the region named after him-Wallacea- the boundary of Indonesian islands that separate Asian fauna from those of Australia and New Guinea.

2.2 Linguistics and Migration Patterns

The first peoples of the tropical Pacific trace their ancestry in waves of migration from Africa through Asia. Archaeological evidence suggests that modern humans were in Australia by about 50,000 years ago, and that New Guinea was occupied later, during the last glacial maximum about 25,000 years ago when sea levels dropped by about 125 m. Human and animal traffic between Australia and New Guinea at that time was facilitated by land bridges that were present for thousands of years, but disappeared after melting of the glaciers and the rising sea levels that accompany them. This gave rise to watery barriers that isolated, indigenous animal populations (Fig. 2.1). A similar effect occurred with pockets of humans in the mountainous terrain of New Guinea in which groups of people and the languages they spoke became isolated.

While Papuan is the tongue unique to New Guinea and adjacent islands, it is not a single language but instead is composed of 23 or more different linguistic families (and more than 900 individual languages) that appear to be unrelated to each other [1]. Of the 3–4 million Papuan speakers, some languages heard in New Guinea are represented by as few as 25–50 people, and are unintelligible even to members of adjacent communities, particularly in isolated mountain villages. Other indigenous cultural groups such as the Huli people (Fig. 2.2) may number 80,000 or more and have lived in the southern highlands for perhaps a thousand years.

Papuan-speaking people are thought to have begun migrating from New Guinea eastward into the Bismarck and Solomon Islands during the lower stands of sea level associated with the maximum extent of Pleistocene glaciation, likely requiring a bamboo raft or dugout to span the relatively short distances. Even now, the distance between Papua New Guinea and the nearest large island in the Bismarck Archipelago is less than 100 km, and there is a smaller island at the halfway mark. By about 3000 years ago Papuan speakers occupied the entire Bismarck and Solomon archipelagoes, a region that spans a distance of about 1500 km and is often referred to as Near Oceania (Fig. 2.3) [1]. From there migrations to island groups that were farther east from the Solomon Islands continued to the island complexes of New Caledonia, Vanuatu Fiji, a region called Remote Oceania.

As Papuan speakers were beginning to explore and colonize Near Oceania, archaeological and linguistic evidence suggests that the first non-Papuan migrants originating from Taiwan were spreading out in new waves of migration to coastal New Guinea and the nearby islands. These people are the likely source of a language family known as *Austronesian* aptly named "southern islands" in Latin and Greek. Austronesian in some form is spoken (but rarely written) by more than 350 million people. There are some 400 languages in that family, and it spans the

Fig. 2.2 The mountainous regions of New Guinea are home to multiple Papuan language groups. Huli is one of the largest with roughly 80,000 speakers. Huli wigmen paint their faces yellow, red, and white and are famous for their tradition of making ornamental wigs made from the hair of family members, as well as their own. The wigs are trussed wide and high with branches, and are often intricately decorated with bird feathers, pig tusks, shells, beads, or flowers (Image courtesy of Wikipedia.org)

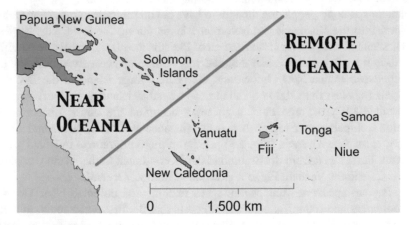

Fig. 2.3 The colonization of Remote Oceania began from Near Oceania about 3000 years ago and required considerable boat building and navigational skills. The two regions are linked by archaeological findings, genetics and language (Modified from Wikipedia.org)

globe from Indonesia and SE Asia to Hawaii and Easter Island in the Pacific, and to Madagascar off the East African coast. Austronesian speakers are now found along the coastal areas of New Guinea, and through the Solomon and other islands of Oceania.

The patterns of development and the geographic expansion of Austronesian are complex and controversial. Depending upon whether the evidence is based on archaeology, culture, linguistics, or genetics, migration dates can vary by several hundred years or more, but two competing models are dominant. The first, called the "out of Taiwan" or "express train to Polynesia" [2, 3] suggests that a rapid migration from Taiwan about 5000 years ago occurred first through the Philippines and Indonesia, then beyond, ultimately extending to Tonga and Samoa. Linguistic analysis [4] as well as archeological information (see below) support the idea of a relatively rapid geographic expansion. However, a different analytical approach that relies primarily on genetics suggests an alternative "slow boat" model where there may have been a roughly 15,000-year period of development within eastern Indonesia before the eastward migration into Fiji and other islands farther east [5]. Regardless of the mechanisms and timing, the inhabitants appear to be a genetic admixture of multiple Austronesian and Papuan groups that intermarried, and interwove their languages and culture as they expanded. Approximate dates of these migratory movements are shown and further described in Fig. 2.6.

2.3   The Lapita Culture

Part of the archaeological evidence for the rapid spread of people into Remote Oceania is the occurrence of stilt houses along with quadrangular stone adzes, distinctive mortar/pestles and domestic pigs. A low-fired, red ceramic pottery, often tempered with shell or sand, was also found in such settlements, often with a distinctive geometric design achieved with a sharp toothed stamp that was pushed into the clay before firing (Fig. 2.4). This earthenware was the hallmark of the Lapita culture that dates to about 3500 years ago. The clay came from the Bismarck Archipelago, and while the designs may have had their origin from there as well, the overall characteristics of the Lapita suggest a Southeast Asian island origin [6]. Over the next few hundred years, radiocarbon dating indicates that their descendants crossed large stretches of ocean and spread south to New Caledonia and to Fiji, Tonga and Samoa in the easternmost part of Remote Oceania [7, 8]. Clay forms from the weathering of silicate minerals, and because atolls are composed of limestone, the material for Lapita pottery must have come from the high islands. However, trade among island groups spread clay pots and related artifacts to adjacent regions. But for reasons unclear, the distinctive pottery itself disappeared. In some regions it may have undergone an evolution to an undecorated ceramic with a limited range of form (plain ware) whereas in others ceramic production seems to have stopped.

Fig. 2.4 Fragment of Lapita pottery from Fiji dated from 3050 years ago (Image courtesy of Patrick Nunn, WikiEducator)

Shifting Cultivation in the Western Pacific

The Bismarck Archipelago is part of the Coral Triangle where the planet exhibits its maximum diversity of fishes, corals, clams and nearly everything else that lives in shallow water. The Triangle holds nearly 600 coral species and 1700 different fishes. In the Bismarck Islands alone there are about 1000 fish species and more than 400 different kinds of corals with reefs covering 80% or more of the available hard bottom substrate [9]. The first colonists would have been quite impressed with what they found underwater. However, that was not the case in the terrestrial environment as far as food was concerned. Birds and fruit bats were abundant, but edible plants were few and far between. Indeed it is apparent that the first permanent settlers brought food staples with them including taro, yams and other root crops, as well as bananas and breadfruit trees described in Chap. 4.

It is hot, humid and rainy on the islands of the tropical western Pacific and this gives rise to heavy tropical forest growth. In order to clear space for crop production, the natural forest had to go, and this was typically accomplished using fire. Crops would be planted and harvested until the soil was exhausted of nutrients. The former forestland would then allowed to remain fallow (uncultivated) for a period of time, usually years, to regain soil fertility by means of microbial activity. In the meantime, new forest land was required. This particular type of slash and burn agriculture is called *shifting cultivation* and resulted in deforestation as well as soil erosion. The latter would have muddied the waters near shore, thus beginning a process of reef degradation, although there is no record of such occurrences. The paradox of shifting cultivation on small islands is that expansion of planting fields for gains in the short term, results in long-term environmental losses. Nonetheless, despite 30,000 years or more of settlement history in the Bismarck and Solomon Islands, and 4000 years of shifting cultivation, they remained heavily forested with secondary growth until the relatively recent advent of clear cutting. These forests regrew on larger islands with the aid of high rainfall and rich volcanic ash, and may have been helped by population controls [10]. Malaria is caused by a protozoan parasite that is transmitted by mosquitoes of the genus *Anopheles*. This disease is endemic to the wet islands of the western Pacific, which extend eastward as far as Vanuatu and may have played a role in keeping human populations low. Ulcerating skin infections, amoebic dysentery and hookworm likely assisted by keeping lives short [8], and perhaps served as motivation to find more salubrious climates.

The Evolution of Naval Architecture

The distances between islands in Remote Oceania are nearly as great as the entire span of the Solomon Archipelago. What may have motivated such long-distance migrations by the Lapita is unclear but it may have coincided with a period of warm weather that produced a wider trade wind belt that decreased the threat of storms [11]. At the same time, these explorers must have developed navigational skills and seaworthy vessels. The evolution of watercraft likely began from rafts that were equipped with a sail to increase speed. However, platforms with sails that merely float on the surface are unstable. A dugout canoe could ride a bit deeper and take advantage of the water's lateral resistance, but a sail posed the same stability problem. A double canoe would have been a better design and may have led to relatively small, single-hulled vessels that were made stable by using an *outrigger*, a second, solid hull lashed parallel to the main body to counteract its tendency to capsize (Fig. 2.5). Although there are no remains of such vessels, it is likely that the Lapita used a dugout log with a single outrigger [12]. However, while there would be a substantial change in canoe technology, it is unclear how they developed and when.

By the time of European contact, ocean-worthy hulls evolved the use of planks that were joined together longitudinally. Holes were drilled and the planks were fastened to each other, and to ribs and keel by stitching or lashing with coconut fiber. Tree sap was used as caulk. There were three basic structural configurations including the single outrigger, the double outrigger, and the double canoe. Of these, the single outrigger was the most widespread in the tropical Pacific region, and their size varied considerably. Smaller outrigger canoes were constructed for shorter-distance travel and some of them used a ballasted, multi-hulled design that was quite sophisticated, size notwithstanding. Sail construction also varied in both size and shape, but was typically woven into a mat, likely coconut or other palm-like plant material sewn together and framed by wood on two sides. A canted rectangular sail was set on a corner and held by a single mast, at least in the area of New Guinea and most of island Southeast Asia. However, in most of the rest of the tropical Pacific, sails were triangular, framed on two sides, and lashed to either the fore or aft of the vessel. Some double canoes used two sails [12, 13].

The most impressive of the long-distance vessels were the Fijian drua. These were much like very large catamarans, hollow double-hulled structures that could carry 200 or more people and were capable of being at sea for months. One hull was larger than the other and some of them were roomy enough for tons of supplies, even a

Fig. 2.5 A single hulled craft with an outrigger such as this Camakau from Fiji was one of many designs used throughout the Pacific islands as voyaging canoes. The framed sail is anchored to the forward part of the canoe and is supported by a central mast. Construction of the considerably larger drua required the replacement of the outrigger with a second hull (Wikipedia.org)

sounder of swine or other livestock. The smaller hull was used in place of an outrigger and was always sailed into the wind. A large triangular sail fashioned from woven palm fronds gave power to these vessels, enabling them to reach speeds of 15 knots or more while still being highly maneuverable, at least while it was under the control of a skilled pilot [14]. Druas and similar ships were very costly to build in terms of time (years), and the size and number of trees that required harvest and transport to build them. Not surprisingly, they were typically the property of royalty.

Expansion into the Central and Eastern Pacific

The islands and environments of the central and eastern Pacific are very different from those in the west. Most are small, including the Society and Marshall archipelagoes, the Tuamotu group, the Cook and Gilbert Islands, the Marquesas group and Easter Island (Fig. 2.6). Rainfall on small islands is highly variable both spatially and temporally, and once outside of the rainy intertropical convergence zones of the western Pacific (Figs. 1.1 and 1.2) rainfall becomes more seasonal and less predictable. The occurrence of tropical storms or El Niño events can change these

Fig. 2.6 The Laysan rail was endemic to Laysan Island in the NW Hawaiian group. No more than a half meter long, these ground-nesting flightless birds were one of many that became victims of human, feral dog and European rat predation (Illustration by JG Keulemans, Courtesy of Wikipedia.org)

patterns, but islands in the central and eastern Pacific tend to be relatively dry, even at the equator. The higher islands may create rain when wind lifts warm air against the slopes of mountains and cools it until it condenses into clouds and rain. This only occurs on the windward side and if the winds are unidirectional, the wet windward side contrasts with a dry leeward side. Of the ten Marquesas Islands, for example, only two reach elevations above about 750 m and these are the only ones that have sufficiently reliable precipitation for the development of small rainforests. Likewise in the Hawaiian Islands there is a considerable diversity of rainfall due to seasonal variation in humidity and the effects of wind-driven uplift. Most of Hawaii is seasonally dry, but on the high islands of Maui, Kauai, the Big Island and Oahu the sides facing the trade winds have a wetter climate.

Islands that receive much less rain and are far from the mainland make them poor places to raise *Anopheles* mosquitoes, thus malaria is rare east of Vanuatu. By contrast, lower rainfall levels also make soil recovery far more difficult to accomplish after it has been used for agriculture. Nonetheless when encountered by Europeans in the eighteenth century, all of these islands were partly deforested; fire-resistant grasses and ferns replaced native trees and shrubs [8, 10]. Many of these islands were able to sustain crops using irrigation and arboriculture (e.g., breadfruit and other trees) as agricultural mechanisms of avoiding shifting cultivation. Easter Island, also referred to as Rapa Nui, is the driest island in the tropical Pacific with a mean precipitation of about 1300 mm per year. However, rainfall is not consistent and as indicated above it can vary from 3 to 4 times that in 1 year, to multiyear droughts in others. When Captain James Cook visited the volcanic island in 1774, he found it treeless and covered with grasses. However, it is known that the island was forested at one time and that a giant endemic palm was dominant within the island's native vegetation. Indeed there are stumps of the palm that remain and many of them are charred, suggesting a purposeful forest removal some 800 years ago [10, 15]. Even on the large Hawaiian Islands, the native forests were often burned to altitudes of at least 500 m and replaced with large agricultural tracts. Palm forests composed of endemic species may have covered the lowlands judging from seed, pollen and trunk remnants, but these had disappeared by the time of European contact [16]. Likewise, in nearly every valley with a permanent stream, the natural vegetation was replaced with irrigated taro and other plants introduced long before the arrival of Europeans [17].

The effects of moving away from the western Pacific are even apparent in the marine environment. A considerable loss of diversity occurs as one moves away from the Coral Triangle. The Marshall Islands for example, has half the number of fish species present in the western Pacific, and the Hawaiian Islands has 20%. By the time lonely Easter Island is reached only about 8% of the fish species of the western Pacific are found although more than 20% of them are endemic [18].

The Lapita and the Plight of Native Birds

As natural forests gave way to agricultural land, many plants that relied on forest growth, including shade plants, arboreal ferns and many vine species for example, disappeared with them. Animals were also affected by this habitat loss, particularly the avifauna. However, forest alteration was not the only factor in their disappearance, reduction of numbers or outright extinction. Native birds were hunted throughout the tropical Pacific, primarily for food. As the Lapita people migrated to islands where birds had never encountered a mammalian predator, they were likely to be easily caught, and there is good evidence that native people, made frequent use of them. Most were hunted by hand or with sticks from trees or even from the sides of steep cliffs, although sticky sap and snares made from saplings were also employed. Birds that had evolved flightlessness or those that nested on the ground were the easiest marks. The flightless rails, perhaps 2000 species of them, are nearly extinct across the Pacific (e.g., Fig. 2.6). Many shearwaters, petrels, albatrosses and other ground nesters were also taken for food. Avian species also had other uses. Colorful feathers were used to create cloaks and headdresses as well as decorations for other objects including weapons (e.g., Fig. 2.2). Red feathers were particularly valued in Polynesia (see below) for their association with major deities. Thus parrots, lorikeets and many other species with red feathers have been extirpated across a wide swath of those islands. Little of them were left unused. Small avian bones were used as needles for sewing and for tattooing, and whistles were fashioned from larger leg bones [19].

It is estimated that within the first millennium of expansion into Polynesia, the native bird population had become substantially diminished. Even though the exact chronology and patterns of extinction are not known with precision, the remains of birds are a substantial part of the archaeological record [20]. It is also worthy to note that humans brought pigs, dogs, and rats with them, and these animals ate ground-nesting birds, chicks, and eggs.

2.4   Characteristics of Melanesia, Polynesia and Micronesia

Melanesia The nineteenth century view of the Pacific islands put their inhabitants into three cultural and ethnic categories. The first of these were named Melanesia, which included islands that extend from New Guinea to Fiji, including the Solomon Islands, Vanuatu and New Caledonia (Fig. 2.7). These were known as the black islands, and they were lumped together primarily on the basis of race. Language, especially but not limited to Papuan and Austronesian, as well as multiple cultural, political, social and geographic influences (described in Chaps. 4 and 5) suggest a complex pattern of migration. Genetic analyses indicate an initial settlement and

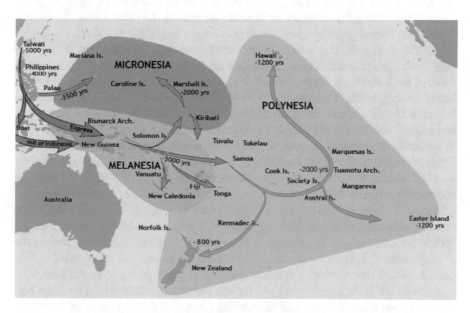

Fig. 2.7 The three classical Pacific island regions are Melanesia, Polynesia and Micronesia. Each has its characteristic history of migration, language development and culture, although the concept of Melanesia as shown and originally described as a racial division is no longer considered valid. 'Express' refers to rapid movement from Taiwan to Melanesia and Polynesia; 'Boat'(slow boat) indicates long pause in eastern Indonesia prior to expansion; negative dates refer to years before present (after [1] Kirch, 2010) and show the expansion into Far Oceania about 3000 years ago. Eastern Polynesia was occupied after a pause of about 1000 years for reasons unclear. Hawaii and Easter Island were occupied no earlier than 1200 years ago and New Zealand appears to have been the last outpost to be occupied during the Polynesian diaspora (Image modified from Wikipedia.org)

expansion in the region about 30–50,000 years ago, with a second important expansion from Island Southeast Asia/Taiwan during the interval roughly 3500–8000 years ago. However, there is also a genetic distinction among the remote and isolated inland Papuan-speaking groups on larger islands, while the coastal groups display a higher degree of genetic diversity and appear to have intermixed among many groups [21]. Therefore most anthropologists do not accept the unitary characterization implied by the term 'Melanesia' or the boundaries it imposes. The distribution of Lapita pottery, for example, indicates an overlap of trade and culture that interwove 'Melanesia' with Samoa and Tonga, societies that are more often depicted as Polynesian. However, the term is used here to designate Melanesia's geographic position in the western tropical Pacific, and its importance in defining the initial expansion and island exploration that began from New Guinea.

Polynesia This region of "many islands" is a large triangular area of Oceania that ranges from Tonga and Samoa on the western side, to Easter Island on the east, Hawaii at the north and New Zealand on the south. Captain James Cook was the first to thoroughly explore this vast region during his three voyages (1768, 1772 and 1776). He and his onboard naturalist were both struck by the similarity of the people and their common culture. Indeed, Polynesia is the only one of the three proposed cultural and biological regions that is relatively homogeneous in terms of genetics, culture and language. Polynesian languages form a well-defined Austronesian subgroup that is divided into western and eastern branches. Interestingly, it has been suggested that the western branch was derived from 'pre-Polynesian', the ancestral form of the Polynesian language that developed on Samoa, Tonga and a few other islands nearby when they were first colonized about 3000 years ago [4]. The eastern branch is associated with the great Polynesian expansion that took place east, north and south from Samoa and Tonga between 2000 and 800 years ago (Fig. 2.7).

In addition to sharing certain of the Austronesian languages and vocabularies, Polynesians have other characteristics in common. In most of the region, especially on high islands, there were only two classes, chiefs and commoners, although on some there were intermediate classes of skilled individuals including artisans, priests, and warriors. The concept of chief could vary considerably. On atolls and small islands the distinction might be minimal or represent a local lineage or decent group. By contrast on larger islands or island groups (e.g., Hawaii, Tonga, the Society Islands) the position of chief was not only inherited, but represented a royal class in which the monarch's spiritual power and prestige, *mana* in Austronesian, came from the gods. Commoners had far less mana, and relationships between persons of unequal mana were *tapu*, from which 'taboo' is derived. Tapu relationships are hardly the sole property of Polynesians. The morganatic marriage of Britain's Edward VII to the American Wallis Simpson was English tapu, and the marriage of Sophie Chotek to Archduke Franz Ferdinand was the Austro-Hungarian version. There were numerous other Polynesian tapu that did not involve marriage. An inferior could not eat food cooked for royalty. A chief's house or others of high rank, as well as their possessions were also tapu for those of lower status. It was even tapu to look at a royal facial tattoo while it was being carved.

Fig. 2.8 Tiki the creator, Marquesas Islands (The Louve, courtesy of Wikipedia)

Other unifying aspects of Polynesian culture include their outrigger canoe technology and navigation techniques that required the motion of specific stars that rise or set on the horizon. Polynesian art forms included common mythological beliefs. Ceremonial human-like statues generally called *Tiki* (Fig. 2.8) were found throughout the South Pacific, although there were variations of the word in Tahitian and Hawaiian. Depending on the island group, Tiki was typically a progenitor god that gave rise to the first man, or at least his penis (Tiki-wanaga), but there were also derivative Tikis. The first fish was Tiki-kapakapa; the first bird was Tiki-tohua, and there was even a first sweet potato, Tiki-whakaeaea. However, in Rarotonga (Cook Islands) Tiki guarded the portals to paradise.

Micronesia This region occupies the region of Oceania north of Melanesia, and extends from Palau and the Mariana Islands on the west, to the Marshall Islands and the Gilbert Islands (now part of the island nation of Kiribati) on the east. The Caroline Islands occupy the center. With the exception of Guam and a few lesser-known volcanic islands, Micronesia is composed of over 2000 small and low coral islands and atolls, hence the name of the region, meaning "small islands" (Fig. 2.9).

Fig. 2.9 Location map of Micronesian islands including the Mariana group in the north, the Carolines (Palau to Kosrae) in the center, and the Marshall and the Gilbert islands in the east. Older island names used here include the Ellice group (now Tuvalu), Truk (now Chuuk), and the New Hebrides Islands (now Vanuatu) (Courtesy of Wikipedia.org)

The language development of the region indicates several migratory events. The first was to Western Micronesia into the Mariana Islands including Guam, and to Palau, and although their language is Austronesian, it is in a subcategory spoken today in the Philippines and Indonesia. This reflects its origin from Southeast Asian islands at least 3000 years ago. Prior to the arrival of the Spanish in Guam, the native Chamorro people constructed unusual elevated buildings using 'latte stones' (Fig. 2.10). These were composed of quarried basalt, limestone or sandstone pillars topped with a hemispherical capstone, often derived from a large coral head. The pillars and were typically spaced in pairs to form a rectangle and likely supported an A-frame type dwelling. It is not clear whether such structures were reserved for the chiefly class or if lattes were typical of a village and status was denoted by pillar height. Chamorro burials were made under or near latte structures but skulls were kept and the spirits of the dead were venerated, but there were no deities worshipped, as was common elsewhere in the Pacific islands [8, 22]. When Guam was declared Spanish territory, mistrust arose between the parties and rebellion ensued. That was followed by harsh and repressive measures to control the Chamorros. In addition, the Spanish introduced diseases such as smallpox and by 1741 the native population was reduced from 100,000 to 5,000 [22].

Fig. 2.10 Latte stones used by the Chamorros to elevate buildings (Courtesy of Hajime Nakano and Wikimedia Commons)

The languages spoken in the rest of Micronesia falls within a distinct cluster of eight Austronesian subgroups referred to as nuclear Micronesian. Linguistic and archaeological evidence indicates that there were several Lapita migrations from Solomon Islands and Vanuatu roughly 2000 years ago [4, 7]. Although Micronesia exhibits clear cultural ties to Polynesia and Melanesia, these have become superimposed on the distinctive cultural traditions and languages that can be traced in some cases to individual island groups. Some common features of the region can be identified including traditional songs, dances and clothing. Households and families as well as land and labor groups are bound together through kinship and matrilineal descent. Thus women in Micronesian society form the core of 'clans', each of which can trace their lineage to the islands first settlers. It is common for clans to group themselves into villages, with grandparents, cousins and associated children, and to distinguish certain social lineages as 'castes' [23]. On the islands of Yap for example, there are different areas or municipalities that are associated with the caste system. The central high islands are home to the highest castes whereas lower castes populated the outer low islands. Outer islanders must wear a simple loincloth and it can be only one color. Main islanders may wear two or three colored layers. Castes participate in dancing competitions, but only among themselves. Lower caste members may dance only when given permission by higher caste members. Women can marry upwardly, but a man who marries beneath his status must assume the role of the lower caste. Even certain foods are caste related.

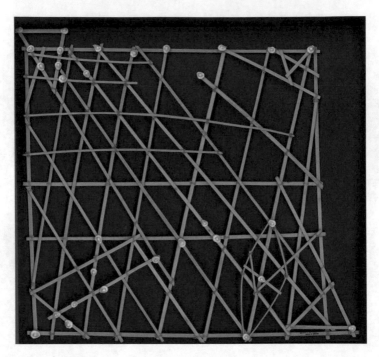

Fig. 2.11 A Marshall Island stick chart showing 30 islands (cowrie shells), predominant swell and counter-swell directions (*curved lines*), and persistent currents (*straight lines*). The islands represented here cover an ocean area of more than two million km^2; Enewetak and Bikini atolls are depicted atop the triangle at upper left, Ebon Atoll is at lower left; Mili Atoll is on the lower right (US Library of Congress)

Some distinctive Micronesian artworks are directly associated with the ocean, especially long ocean voyages. Marshall islanders designed stick charts for navigation and typically depicted swell patterns and currents using coconut fronds tied together with fiber; cowrie shells represented island locations (Fig. 2.11). Weather charms from the Caroline Islands, known as hos, were stylized human figures that were thrust into the wind by the navigator, and when accompanied by the proper incantations, were thought to prevent or divert approaching storms [24]. Caroline islanders developed a sophisticated star compass system based on the rising and setting of stars and constellations. Any given star always rises from the same point on the eastern horizon and sets in the same on the west, thus forming an east-west 'starpath'. When a star has risen far above the horizon, several others will take its place. The Caroline Islands are situated close to the equator where the star paths change very little compared with less tropical regions where the opposite is true. Thus the Caroline star compass was capable of defining 32 distinct directions that could be followed from one island to another and back again [25]. Experienced sailors may have known that land was near in spite of not being able to see it. Clouds tend to form from warmer rising air that occurs on land during the day, compared with the less variable surrounding water temperature. Swells at sea may be interrupted or refracted in patterns that indicate the presence of an island [26]. Roosting birds leave island homes in the morning to forage and return to nest sites at dusk, whereas bats are nocturnally active and display the opposite pattern.

References

1. Kirch PV (2010) Peopling of the Pacific: a holistic anthropological perspective. Annu Rev Anthropol 39:131–148
2. Diamond JM (2000) Taiwan's gift to the world. Nature 403:709–710. https://doi.org/10.1038/35001685
3. Holden C (2008) Polynesians took the express train through Melanesia to the Pacific. Science 319:270
4. Gray RD, Drummond AJ, Greenhill SJ (2009) Language phylogenies reveal expansion pulses and pauses in Pacific settlement. Science 323:479–483. https://doi.org/10.1126/science.1166858
5. Oppenheimer SJ, Richards M (2001) Polynesian origins: slow boat to Melanesia? Nature 410:166–167
6. Denoon D (1997) The Cambridge history of the Pacific Islanders. Cambridge University Press, Cambridge
7. Denham T, Ramsey CB, Specht J (2012) Dating the appearance of Lapita pottery in the Bismarck Archipelago and its dispersal to remote Oceania. Archaeol Ocean 47:39–46
8. Kirch PV (2000) On the road of the winds, an archaeological history of the Pacific Islands before European contact. University of California Press, Berkeley
9. Hamilton R, Green A, Almany J (2006) Rapid ecological assessment- Northern Bismarck Sea, Papua New Guinea. The Nature Conservancy Report 1/09. http://pdf.usaid.gov/pdf_docs/Pnadp130.pdf
10. Rolett BV (2007) Avoiding collapse: pre-European sustainability on Pacific Islands. Quat Int 184:4–10
11. Nunn PD (2007) Climate, environment and society in the Pacific during the last millennium. Elsevier, Amsterdam
12. Irwin G (2008) Pacific seascapes, canoe performance, and a review of Lapita voyaging with regard to theories of migration. Asian Perspect 47:12–27
13. Finney B (2007) Ocean sailing canoes. In: Howe KR (ed) Vaka Moana voyages of the ancestors-the discovery and settlement of the Pacific. University of Hawaii Press, Honolulu
14. Nuttall P, D'Arcy P, Colin P (2014) Waqa Tabu- scared ships: the Fijian drua. Int J Maritime Hist 26:427–450
15. Mann D, Edwards J et al (2008) Drought, vegetation change, and human history on Rapa Nui (Isla de Pascua, Easter Island). Quat Res 69:16–28
16. Chapin MH, Wood KR et al (2004) A review of the conservation status of the endemic Pritchardia Palms of Hawaii. Oryx 38:273–281
17. Cuddihy LW, Stone CP (1990) Alteration of native Hawaiian vegetation: effects of humans, their activities and introductions. University of Hawaii Press, Honolulu
18. Randall JE, Cea A (2011) Shore fishes of Easter Island. University of Hawaii Press, Honolulu
19. Steadman DW (1997) Extinctions of Polynesian birds: reciprocal impacts of birds and people. In: Kirch PV, Hunt TL (eds) Historical ecology of the Pacific Islands. Yale University Press, New Haven
20. Steadman DW (1999) The prehistoric extinction of South Pacific birds. In: Galipaud J-C, Lilley I (eds) The Pacific from 5000 to 2000 BP Institut pour le Développement, Paris, pp 375–386
21. Friedlaender JS, Friedlaender FR et al (2007) Melanesian mtDNA complexity. PLoS One 2(2):e248. https://doi.org/10.1371/journal.pone.0000248

22. Rogers RF (1995) Destiny's landfall, a history of Guam. University of Hawaii Press, Honolulu
23. Petersen G (2009) Traditional Micronesian societies: adaptation, integration, and political organization in the central Pacific. University of Hawaii Press, Honolulu
24. Kjellgren E (2014) How to read oceanic art. Yale University Press, New Haven
25. Hutchins E (2014) Understanding Micronesian navigation. In: Gentner G, Stevens AL (eds) Mental models. Psychology Press, New York, pp 191–225
26. Irwin G (1989) Against, across and down the wind: a case for the systematic exploration of the remote Pacific islands. J Polyn Soc 98:167–206

Chapter 3
European Exploration of the Pacific During the Age of Discovery

Abstract The exploration of the Pacific began with Portuguese interests in the Indian Ocean and adjacent seas. This chapter begins with the establishment of trading posts in Ceylon, India and the Maldive Islands, and eventually in the Spice Islands of Indonesia, all of which had already been heavily influenced by traders from the Middle East. In the late fifteenth century the eastward traffic from Europe used the Indian Ocean route, pioneered by Vasco de Gama, a pathway that never actually crossed the ocean. In an effort to find an alternative route to the Spice Islands, Ferdinand Magellan took the first epic voyage across the Pacific in 1520. He discovered and described the low islands he encountered there, but there would be no renewal of food or water until reaching the Mariana Islands. Here he would encounter the Chamorro people on the island of Guam after being at sea for 98 days. The interaction would become a sociological template for future dealings between native Pacific people and the Europeans, as evidenced by the more extensive Pacific explorations made by Alvaro de Mandaña de Neira and Pedro Fernandez de Quirós beginning in 1567. The discoveries and difficulties associated with those three voyages are described in detail.

3.1   The Portuguese, the Spice Trade and the Prelude to Pacific Exploration

The late fouteenth and early fifteenth centuries are referred to in Europe as the Age of Discovery, a time of global exploration that is sometimes regarded as a cultural bridge between the Middle Ages and the Age of Reason. European seamen were a rough lot, and they had to be. The pay was low, the journeys were long, and the food was what could survive without refrigeration, including salted beef or pork, weevil-infested hardtack, and dried beans or peas. After several weeks at sea, rat droppings contaminated much of it. Ale or hard liquor was available, but the demand was high and it typically ran out early. There were no fruits or vegetables that could have reduced scurvy, the scourge of the long voyage (see Text Box). Some of the crew were seasoned veterans, others were very young and allegedly met the minimum age of 16 years. There were those who were petty criminals, and some who were forced into service. There was oftentimes friction between the crew and the officers;

© Springer International Publishing AG 2018
W.M. Goldberg, *The Geography, Nature and History of the Tropical Pacific and its Islands*, World Regional Geography Book Series,
https://doi.org/10.1007/978-3-319-69532-7_3

beatings and flogging were commonly employed for minor infractions as a means of keeping the rowdiest seamen in line.

First the Portuguese, then the Spanish, and finally in the late fifteenth and early sixteenth centuries, the British, French and Dutch left the waters of the Old World and embarked on their adventure into the vast "green sea of darkness" to discover new land, new goods and new trading partners.

Text Box: The Purpurea Nautica

The role of vitamins in nutrition was unknown and would remain a mystery until the 1930s when some of them were discovered, including vitamin C. This nutrient had a particularly important role in maintaining connective tissue including that found in skin, muscle, veins and joints. Certain fruits and vegetables were rich sources of it, but such foods were not available on long sea voyages because they were prone to spoilage. After a few weeks sailors developed what was called 'purpurea nautica' for the purplish bruises that came from fragile and broken capillaries (Fig. 3.1). That served as the first indication of the disease that would become better known as scurvy. Lethargy due to anemia often accompanied the bruising. After six weeks or more, scurvy began to manifest itself as swollen and bleeding gums that resulted in loss of

Fig. 3.1 One of the first signs of scurvy, the purpurea nautica, due to fragile capillaries in the skin (Courtesy of James Heilman and Wikipedia)

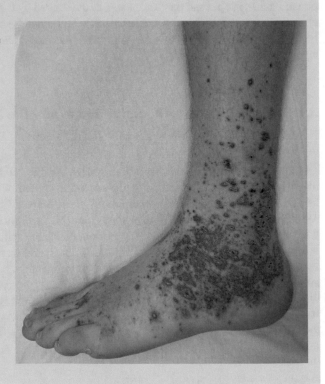

teeth, wounds that would not heal, internal hemorrhaging and degeneration of the joints, especially those of the legs. Jaundice, a liver condition that is accompanied by yellowing of the skin and eyes, was also a common symptom. Death would eventually result from infection or blood loss. Early voyagers were at a loss to explain the disease and blamed it on bad or 'pestilential air', water or putrefied food, the same causes often ascribed to those who developed yellow fever, cholera and other illnesses caused by viruses or bacteria.

Attempts to treat scurvy with yeast or fresh fruits and vegetables began in the mid-18th century. Captain James Cook carried vitamin C-rich sauerkraut aboard ship during his first voyage to Tahiti in 1768 and stopped at every port for fresh fruits and vegetables in an attempt to test their efficacy in combatting the disease, even though such treatments were considered anecdotal [1]. It would not be until the end of the century that the Royal Navy routinely issued lemons or limes to sailors on long voyages. Thus was born the derogatory term for His Majesty's sailors, the limeys.

Global maps pinpointed far-flung locations with increasing accuracy, as European powers became the masters of vast colonial empires. With increasing contact between the Old and New Worlds, the Columbian Exchange began to take shape: a massive shift and transfer between west and east, of animals especially livestock, and plants, particularly agricultural products. There were also exchanges that tended to be one way, including the enslavement of native peoples and the export of microbes including smallpox, typhus, influenza and measles to the Americas and eventually into the Indo-Pacific.

The Portuguese explorer Vasco de Gama was the first European to reach southern India by sea in 1498. In this effort he left Lisbon, rounded South Africa and hugged the east African coastline until reaching SE Kenya. From there he crossed the Arabian Sea to reach SW India, taking 23 days to complete the journey in the open ocean. Here they found a welcoming Hindu population, a religion de Gama had never heard of, and established a presence in an effort to establish a trade in spices. They began to explore islands from there, particularly Ceylon (Sri Lanka) and the nearby Maldives, an archipelago off the southwest coast. According to The Codex Hispanus in about the year 1500, there were 12,000 Maldive Islands large and small. Some were five leagues apart; others the distance of a crossbow shot, and a few could be reached on foot at low tide. The men and women of these islands were all black but not as black as the people of Guinea, and their hair was straight, and not frizzy like the hair of Ethiopians. They were rustic but simple and clean living, and they had their own language [2].

In the Indo-Aryan Divehi language spoken in the Maldive Islands, mal means a thousand, and diva is the term for island. Official estimates put the number at 1190 and thus the name of the island group is close to being literally correct, even though

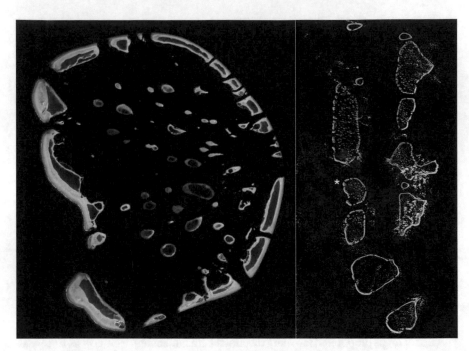

Fig. 3.2 *Right*: Atolls of the Maldive Islands seen from space. North Nilandhoo position shown by *asterisk* (NASA photo). Note that the lower two atolls and others not shown are without prominent faroes as shown by North Nilandhoo Atoll at *left*. Here, faroes compose the rim and are prominent features of the lagoon. This atoll is 26 km across at the center (Courtesy of Landsat, US Geological Survey)

the Portuguese inflated it by a factor of 10. Regardless of the exact number, only 21 are atolls and the origin of that word comes from the Divehi, atholhu, referring to the ring-like shape. However, the Maldivian atolls are a bit out of the ordinary. Lying on a north-south volcanic ridge, the Maldives reefs began growing toward the surface about 50 million years ago, but were interrupted by sea level fluctuation and erosion as in the Pacific. There is a considerable amount of variation among them, but in the northern group the rims and even the lagoons are studded with miniature atolls called *faroes*. These structures emerge at low tide and may have diameters of 10 to more than 100 m, while their miniature lagoons may be 20 m deep or more (Fig. 3.2). Each one is often separated from the others by deep channels. Faroes become less prominent toward the south in the Maldives for reasons that are unclear. Darwin thought the change might have been due to greater subsidence from north to south, and indeed lagoons deepen away from the equator. However, it could also be due to the underlying erosional features of the platform, the influence of reversing monsoon winds, and the frequency of cyclones that also increase in the same direction.

These islands were in fact described prior to the arrival of the Portuguese in the Ceylonese Buddhist literature, which represented the predominant religion on the

Fig. 3.3 Typical 2–3 cm shells of the money cowrie, *Monetaria moneta* (Photo by author)

Maldives until the mid-late twelfth century. By that time conversion to Islam had begun under the influence of Middle Eastern merchants who became interested in the islands as way station that linked their trade routes to eastern Africa. Several Arab writers describe various aspects of life in the Maldives, but none as extensively as the Moroccan explorer Muhammad Ibn Batutta, who lived there for nine months in 1345 and described two products for which the islands subsequently became well known. The first was coir, fiber extracted from coconut husks that Maldivians fashioned into ropes. This cordage was vastly superior to hemp in strength and in resistance to rot. Soon it became an export to India and was the standard for ship ropes and rigging throughout the region [3].

The second item was a cowrie, small mollusks whose shells were used as a medium of exchange in certain parts of the world and were used in ornamentation or worked into beads [4]. There is a specific "money cowrie" (*Monetaria moneta*, Fig. 3.3) that was particularly abundant in Indian Ocean lagoon environments. They were typically about 2–3 cm long or less, but several similar species were used for commodity exchange between places where they were common, and in places such as Africa and Asia where they were not. Unknown to the early explorers however, the same cowries were found throughout the tropical Pacific, where they were widely used in a similar manner.

They were so abundant in lagoons of the Maldives that millions could be gathered fairly quickly and vessels would often carry them as cargo to Indian markets for trade, including the Portuguese colony that had been established in SW India, then known as Malabar Coast. However, the Portuguese wanted to be certain that Indian Moorish influence around the region was weakened, and to that end a trade ban was implemented on the importation of money cowries among other items of commerce. Vasco da Gama had just departed for Lisbon, leaving India and a squadron of five ships, the remainder of the Indian Ocean fleet to his uncle, Captain Vicente Sodré. During a patrol in 1503 Sodré discovered four Maldivian vessels loaded with cowries destined for Malabar in defiance of the ban. The captain then demanded that the Maldivians point out the Moorish merchants and after taking the cargo from the ships, the merchants were bound, forced into the hold of one vessel, and the ship with its human cargo was set ablaze [4]. A few years later, the Portuguese signed a trade agreement with the Maldives after making several attempts to simply take them over. They eventually succeeded in 1558, but ruled the islands for only 15 years before being driven out. The Portuguese tried twice more in the seventeenth century to regain the islands, but by then the King of the Maldives had used his supply of cowrie shells to forge a relationship with the Dutch Burghers in Ceylon in exchange for protection [4]. Meanwhile, the cowrie from the Maldives had become so established as a form of exchange by the Portuguese that they were used in the African slave trade. In 1520 for example, a male slave destined for the Atlantic-African island of Sao Tomé fetched 6000 cowries [5, 6].

3.2 Magellan's Pacific Odyssey

Having established an outpost by force in 1511 on what is now the island of Sumatra, the Portuguese were very much interested in exploiting nutmeg, pepper, cinnamon, ginger and similar commodities in East Indies, now Indonesia. There was fierce competition for these items, especially between Portugal and Spain, and thus the most expeditious route to these islands was sought. The Portuguese explorer Ferdinand Magellan (Fernão de Magalhães in Portuguese) wanted to seek a trans-Pacific westward passage, but king Manual of Portugal was uninterested and rebuffed several attempts by Magellan for funding. Frustrated, he moved to Spain where he married into an influential family and was granted an audience with the Spanish King Charles who outfitted Magellan in 1519 with five ships and 270 men to claim glory for his adopted country. This Magellan became the first European tasked with making the journey across the Pacific.

After emerging from the South American straits that now bears his name, Magellan entered the Pacific Ocean, which he named Mar Pacifico (Portuguese: tranquil sea) for the unusually calm conditions he encountered. He sailed for more than eight weeks until reaching land, likely the eastern portion of the Tuamotu Archipelago in what is now French Polynesia (Fig. 1.14). Nearly all of the islands here are atolls, small, low, sandy, and typically without fresh water. The one Magellan landed on in January 1521 was located at "13 or 14 degrees" [7] and what it lacked in human inhabitants, was more than compensated for by large numbers of sharks. Because the island had no accessible anchorage, the men would have been required to jump into the water from their longboats to get close to shore. However, all of those dorsal fins induced the crew to return to the ship and continue without a stop at what they appropriately called Isla Tiburones (Island of Sharks). Twelve hundred km later Magellan encountered two more islands, also likely atolls, but like the first one there is no certainty of where they were even after a computer analysis nearly 500 years after the voyage [8]. Although these two islands were hundreds of kilometers apart, they held in common the lack of exploitable resources and were lumped together as "the Unfortunate Isles" [9, 10]. Although Magellan may have discovered the Tuamotu Archipelago, he had little to say about them. Good weather notwithstanding, his men had been at sea for three months and twenty days after leaving the passage out of South America. Their food had all but run out and the men who survived were suffering from malnourishment, dehydration and scurvy.

Eventually, Magellan reached the Mariana Islands in the western Pacific, including Guam, the largest of the island chain. Here they would meet the indigenous Chamorro people (Chap. 2) who greeted Magellan and his crew excitedly but their communal concept of ownership led to looting anything that was not nailed down and the region was dubbed "Insulas Ladrones" (Islands of Thieves). After driving away their guests with crossbows, Magellan's men made their way to the island where in apparent revenge for the thievery, they set a village on fire and killed eight Chamorros who resisted [11]. Even so, and by circumstances that are unclear, on the following day the Spanish were able to trade iron goods for desperately needed food and water before sailing to the Philippines. After arrival on the island of Cebu, Magellan decided that it would be a blessing to convert the natives to Christianity. One local chief consented, and convinced Magellan to join them in their fight against their unconverted enemies on a neighboring island. The captain agreed, much to the disagreement and chagrin of his men. Magellan led the attack, assuming his European weapons would ensure a quick victory, but the unconverted fought back fiercely, especially after seeing their houses set afire by Magellan's men. After being wounded in the arm and leg by a hail of bamboo spears, Magellan was set upon by clubs and spears and was then dismembered with long knives where he fell on April 27, 1521. Only one ship of the original five and 18 men of the original 270 survived the voyage. Among them was Antonio Pigafetta, a scholar who had kept a detailed diary of the expedition [7].

3.3   The Spanish and the Pacific Voyages of Pedro Fernandez de Quirós

La Primera y Segunda Introducción

Other Portuguese explorers followed Magellan's northwesterly route across the Pacific and discoveries were made, albeit by chance. Diogo da Rocha is credited with discovering some of the high islands of the Caroline group after being blown off course in 1525. Alonso de Salazar is credited with sighting the atolls of the Marshall Islands in 1526, although he never explored or landed on them. However, after establishing colonies in Peru, the Spanish embarked on different trans-Pacific routes from the port of Callao, rather than going all the way round South America. Alvaro de Mandaña de Neira launched first such cruise in 1567, and Mandaña and his chief pilot Pedro Fernandez de Quirós began a second one in 1595. De Quirós alone captained the third cruise in 1605, following Mandaña's death. All three voyages had similar goals that included the exploitation of gold, spices and other commodities, the conversion of the native people to Christianity, and the search for the unknown continent of Terra Australis. The latter was a hypothetical continent derived from the Aristotelian deduction that Earth with its vast Arctic-temperate region required balance from an equally massive and expansive southern landmass. Such as continent was depicted on maps for hundreds of years despite any proof of its existence, and indeed it was suggested that the terra incognita must extend very close to the islands of Polynesia. How else could the indigenous people have gotten to such far flung places?

Mandaña's first cruise was successful in discovering one of the larger volcanic islands of the Solomon group east of New Guinea, although he believed that this was Terra Australis. Nonetheless, excitement began to build on the apocryphal tales of riches to be had there, and thus the islands were named after the biblical mines of King Solomon years later. However, even without the gold the islands were rich in desperately needed food that the islanders were able to supply. Unfortunately, bartering for comestibles by the crew led to cycles of friendly welcome, followed by misunderstandings, robberies, and violent reactions that often became the trademark of such cruises. Mendaña returned to Peru losing a third of his 150-man crew but reported that the islands were well suited for agriculture and were useful hunting grounds for slaves to be used in other Spanish territories. They also charted the approximate location of certain atolls in Tuvalu, the Marshall Islands, and lonely Wake Atoll located 3700 km west of Honolulu [12, 13]. There was no rush to further explore these places.

The second voyage, captained by Mandaña in 1595, was much larger than the previous one, with four ships and 378 men, women and children brought for the express purpose of colonizing the Solomon Islands from Peru. Of the men, 200 were capable of bearing arms, including front-loading musket predecessors called arquebuses. Among the potential settlers were a good number of criminals, prostitutes and adventurers, whose presence perhaps led to some of the unfortunate incidents. Here is a brief account of the first, which took place at the first landfall in the high islands of the Marquesas group in the eastern Pacific.

Four hundred of their people came out in canoes, people 'almost white, and of very graceful shape', one youth so clear and fresh and beautiful said de Quirós, that 'never in my life have I felt such pain as when I thought that so fair a creature should be left to go to perdition.' All seemed to be going well for a while, but when the natives began helping themselves to the ship's gear and food, signs were made for them to go, but they were eager not to leave. A gun was fired, and when the Marquesans felt and heard it they swam to their canoes. However, one stubbornly remained clinging onto the ship by its rigging, and he would not let go, until a crew member stabbed his hand with a sword. His fellow islanders then brandished spears, threw stones, and tried to tow the ship ashore. That is when the shooting began.

Fig. 3.4 Going Postal: an ironic philatelic dedication to de Quirós' discovery of the Marquesas Islands of French Polynesia where his crew killed about 200 Marquesans 'mostly for sport'

When the Spaniards left only two weeks later, Mandaña's chief pilot, Pedro Fernandez de Quirós estimated that 200 Marquesans had been killed, mostly for sport [12]. And this was the first substantial contact between Europeans and Polynesians (Fig. 3.4).

The voyage continued west, but instead of reaching the main cluster of the Solomon Islands, Mendaña discovered the more remote Santa Cruz Islands about 400 km to the southeast, and decided that was close enough. However, there were issues with the establishment of the colony. First, Mendaña's leadership abilities were questionable and resulted in continual bickering among the officers and crew. Secondly, the islanders, who were not always cooperative, were massacred and their mutilated corpses were hung in prominent places to encourage harmony, but that inducement was insufficient. Raiding local gardens could not sustain the colonists, and they were not very adept at raising food on their own. Malnourishment set in and was accompanied by malaria and perhaps typhoid fever. Mandaña himself died there of a 'fever epidemic' and the attempted settlement failed. Pilot de Quirós is credited with bringing his vessel to the Philippines (even though the charts he had did not show those islands), but another frigate that carried Mandaña's body disappeared. Fewer than 100 of the original 378 colonists survived the return to Acapulco, Mexico from the Philippines 10 months later [12, 14].

The Third and Final Voyage

De Quirós, now the captain, left the port of Callao in 1606 with his fleet of three vessels, 130 men, the usual armaments, and food for one year. There were also six bare-foot Franciscan friars brought along as models of the faithful life, and four brothers of Juan de Dios, a Catholic order whose mission was to cure the sick [12]. As the three vessels navigated from the coast of Peru to discover the unknown regions of the southern Pacific for His Majesty, de Quirós laid down the law to promote good order, and to provide guidance gained from previous experience with native peoples of the Pacific. It should be remembered that both Mandaña and de Quirós had the same general objectives as on the previous voyages: discover new lands, describe and exploit resources, and save heathen souls. Religious rules were the first to be published, followed by rules of life aboard ship, but the very detailed list of rules for dealing with the native people is the most instructive.

Text Box: Voyage Rules

The captain demands of the crew that they must not curse nor blaspheme, nor say or do other things evil against God our Lord, nor against the most holy Mother, nor against angels, saints, or things divine or sacred. In the afternoon, all must go on their knees before an altar where there are images of Christ and of the Virgin Mary, praying that God our Lord may guide us and show us the lands and people we seek.

Further, I charge him not to consent to any playings with dice or cards, either for small or great stakes.

He is also to be careful to note flights of birds, shoals of fishes, and signs of land, and whether inhabited or uninhabited, place them on the chart in their latitude, longitude, and form, as well as where wood and water can be obtained, as well as the rocks and reefs that are met with.

The colour, shape, features, and dress of the inhabitants are to be noted, their food, arms, boats, behaviour, and government and religion. They must see that the provisions do not turn bad and are not wasted.

After anchoring in any port, a careful look-out should be kept both by day and night, for the natives are great swimmers and divers, and might wedge up the rudder, cut the hawsers, or set fire to the ship.

Take care not to allow so many natives on board the ship as would be able to overpower the crew; and even when they are few, great evil may come to them as well as to us, from ignorance of our arms; whence may arise a commencement of war, and a faithful peace may never then be made.

In effecting a landing, it should always be by day, and never at night. The landing-place should be level and clear of woods, with arms ready, marching together and in order, and entering passes with caution. It should be kept in mind that the natives usually get behind rocks or trees, or stretch themselves flat on the ground even in level places, concealed only by the grass.

Take notice that, if it is possible, chiefs or other natives who appear to be of consequence, should be kept in the ship as hostages, but well treated and given presents of things that they seem to like most. The barter should be conducted by one of us, who should always give the natives to understand that the things are of great value, because they do not value their own things much, and ours but little.

Learn from the natives whether there are other islands or extensive lands near, if they are inhabited, of what colour are the natives, whether they eat human flesh, if they are friendly or carry on war. Enquire whether they have gold in dust, or in small lumps, or in ornaments; silver worked or to be worked; metals, all kinds of pearls, spices and salt, and if they eat those commodities. If they have names for them, write the names down. Ask in what parts these things are to be found, and what those lands are called.

Do not think little of the natives, for they are pilferers and runners. Do not follow the guidance of the natives except with great caution. Never trust or believe in them on any occasion whether they show much or little sign of friendship. They are capable of leading those they pretend to guide direct to their traps or ambushes, or to get them away from their boats or the beach, and to lead them inland into the woods, and there do what evil they can to them.

Never allow our people to mix with the natives, nor leave them to join company.

On occasions when it is desirable to have an interview with the natives, it should always be in a cleared space, with a good distance between the two parties, and the Chief, or one named by him, standing in the space, so as to concert with him what they desire or ask for. It is always necessary to see that the back is safe without ceasing to watch or even turning the face, but always the whole body. And, when obliged, let it be back to back, with the shields in front, so as to make all more strong and secure.

The natives never give up anything they have about them, or anything in their houses, though it be gold, silver, pearls, or any other thing of value, nor do they understand our covetousness. But [we should teach] them to sow maize, beans, onions, cotton even on desert islands. If the place is suitable, rabbits, goats, and swine should be landed, for it is an advantage to enrich those desert lands, remembering the possible needs of future navigators.

Take care not to feed on the things which the natives present to be eaten, because they know how to play tricks. For which reason do not fill your hands, nor quit your arms, nor take your eyes off the natives.

Care should be taken to look out for poison put into the water or food. Vegetables and fruits should not be eaten unless known before, or unless they have been used as food by birds and monkeys [13].

A First Sighting of Islands

This voyage had some difficulties from the start. His Spanish crew disrespected de Quirós as a navigator and otherwise failed to put much trust in him, in part due to his Portuguese heritage and his indecisiveness. After sailing for more than five weeks they encountered their first island in the Pitcairn group of the eastern Pacific (Fig. 3.5). Anchoring was impossible and assuming there were any provisions to be had, no landing could be made. They sailed on for another three days and encountered a second island, long and flat, but very green and full of trees and open spaces. But even though a skiff was able to get immediately next to where the surf was breaking, the bottom was still more than 50 m deep. The coast was completely enveloped in cliffs, and there was no place to land, except for the gulls and terns. They initially logged it as "Sin Puerto" (No Port), but this was isolated Henderson Island, and with an elevation of 30 m it is a good example of an uplifted atoll. It remains remote, uninhabited, and without freshwater even today, and for de Quirós in 1606 it was not worth the risk of getting ashore, even if they could.

The encounter with Henderson Island was thus described by de Quirós: It was a dark, confused, ugly, and long night which we passed, confiding, after God, in the soundness of our ships and the stoutness of our sailors. When the long-wished-for daylight came, we saw that our land was an island surrounded by a reef. Neither port nor bottom could be found, though sought for with care, as we were in want of water, and for fuel we only had brushwood. Seeing that the island was so useless, we left it for what it was; and, considering the night it had given us, it would have been dear even if it had been a very good land instead of a very bad one [13]. They pressed on.

Other bits of land were sighted but they were string of uninhabited low, flat islands so narrow that "there was not a stone's throw from one side (the sea side) to the other". The other sides in these cases were lagoons that might have offered some anchorage and safety, but the pilot failed to see any entrances through the reef. They had entered the south easternmost part of the Tuamotu Archipelago, a spectacular group of atolls, but certainly no Garden of Eden, and rather unhelpful to sailors running low on provisions.

It had been more than seven weeks since leaving Peru. The tenth land sighted on February 10th might just be the charm. There was a rising plume of smoke, which almost certainly meant there were people on this island, and supplies to be had. De Quirós named this island Conversion of San Pablo for the date of arrival on the Catholic calendar. It had an elongated, oval shape and a prominent point with an island extending southeast that was full of palm trees. Navigator Luis Vaes de Torres gave his position on that day as 18°30′, and chief pilot Gaspar Gonzalez de Leza gave it as 18°10′ making it reasonable to suspect that this was Hao Atoll [12], as indicated in Fig. 3.5, but no one is quite certain which island it was. Landing at Hao was just as difficult as on the other islands, but the ship's crew was so excited on seeing the great desire of the natives to see them, the sailors abandoned the long-boats and swam for shore. Perhaps surprisingly, the interactions between crew and

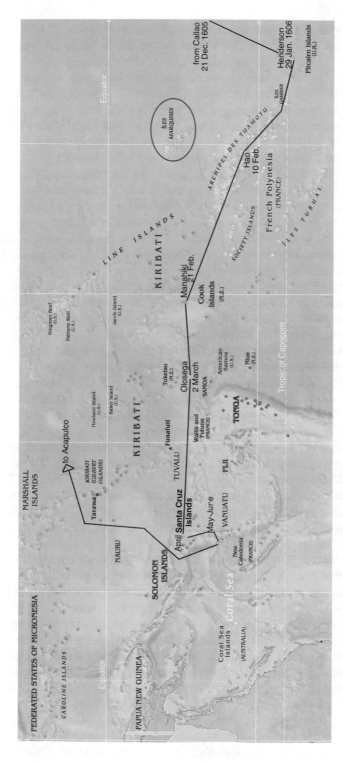

Fig. 3.5 The second trans-Pacific voyage led by Mandaña with de Quirós as chief pilot. The high islands of the Marquesas (circled in red) was the first stop. The route taken during the third voyage (1605–1606) was led by de Quirós, departing from Callao, traversing the Tuamotu group with stops as noted in the text. Dates and general aspects of the route taken are taken from Markham 1904 [13] (Map courtesy of the Perry-Casteñeda collection, University of Texas Libraries)

the natives went well. Gifts were exchanged and no weapons were used, but neither food nor water was obtained. De Quirós wrote about this in his journal and also emphasized a certain attitude that he had brought with him to this latitude.

As soon as they landed the natives, putting down their lances, all together at one time bowed their heads and arms, and saluted three times. Apparently, the welcome and smiles were to receive our men, and when one was knocked over by a wave, they picked him up, embraced him, and kissed him on his cheeks, which is a way of showing friendship used also in France. When the people in the boats saw the loyalty with which the natives received complete strangers, not knowing their intentions, two others went on shore. Marching in good order, they entered a palm grove, where they found, at the foot of a tree, a number of brown stones, and one in the form of an altar, covered with branches. It was supposed that this was a burial-place, or a place where the Devil spoke to and deceived these miserable natives, without there being any one to obstruct him. Our people, to sanctify the place, set up a cross, and gave God thanks on their knees for being the first to hoist His royal standard in an unknown place inhabited by heathens [13].

The counting and naming of islands continued through February 14th, 1606, but none provided anchorage and from the brief descriptions of them, the route through the Tuamotu Islands remains difficult to determine with any certainty. Thinking that de Quirós had rediscovered San Bernardo Island that he had charted in 1567, that was the name given to land sighting #14, although that was a mistake. That island was most likely the easternmost of the Cook Islands group, (Fig. 3.5). It was positioned at 10°40′ S, and was about 10 leagues circumference (55 km) with a shallow anchorage at the north. These characteristics strongly suggest that de Quirós had found Manihiki Atoll [7]. As it was otherwise uninhabited, Manihiki became the Island of Fish, aptly named, as it was time to restock.

As described by de Quirós: There was a great number of fish inshore, and, owing to the water being very shallow, they were killed with swords and poles. There were great numbers of lobster and craw-fish, and other kinds of marine animals. They found a great quantity of cocoa-nuts in a heap at the foot of the palm trees, many large, and of different sizes. There were a great quantity of sea birds of several kinds, and so importunate that they seemed to want to attack the men. We took plenty of all these things [13].

They pressed westward for an additional twelve days to an island they called Peregrino, identified later as Olosega Atoll, the northernmost outpost of American Samoa with 60 hectares of land and a lagoon closed off from the sea. It was inhabited at that time, and the lagoon may have had a layer of fresh water on the surface that was recharged by rain. Captain de Quirós sent an armed party on shore to get sorely needed wood and water, and to describe the native houses and implements.

De Quirós described the Olosegan islanders as follows: In the houses of the natives a great quantity of soft and very fine mats were found, and others larger and coarser; also tresses of very golden hair, and delicate and finely woven bands, some black, others red and grey; fine cords, strong and soft, which seemed of better flax than ours, and many mother-o'-pearl shells, one as large as an ordinary plate. Of these and other smaller shells they make, as was seen and collected here, knives, saws, chisels, punches, gouges, gimlets, and fish-hooks. Needles to sew their clothes and sails are made of the bones of some animal, also the adzes with which they dress timber. They found many dried oysters strung together, and in some for eating there were small pearls.

The land is divided among many owners, and is planted with certain roots, which must form their bread. All the rest is a large and thick palm grove, which is the chief sustenance of the natives. Of the wood and leaves they build and roof their houses [in which] all the floors covered and lined with mats, also made of palms; and of the more tender shoots they weave fine cloths, with which the men cover their loins, and the women their whole bodies. Of these palms the natives also make their canoes, and some very large vessels, twenty yards in length and two wide, more or less, in which they navigate for great distances. They hold about fifty persons. Their build is strange, there being two concave boats about a fathom apart, with many battens and cords firmly securing them together [description of a catamaran]. Of these palms they make masts, and all their rigging, sails, rudders, oars, paddles, utensils for baling, their lances and clubs. On these palms grow the cocoa-nuts, which serve them for food and drink, grease for their wounds, and cups to hold their water. It may almost be said that these trees sustain the good people who are here, and will remain in the wilderness until God takes pity on them [13].

"What we have here is a failure to communicate" Cool Hand Luke

At first, nearly 100 islanders paddled out from their village, singing in unison with each stroke. Two boats and 60 men from the Spanish vessels were placed in the water to meet them. They made hand gestures to the ship's launch indicating that they wished to be followed back to the island, and when that did not happen, they attempted to pull the anchor and tow the boat ashore on their own. The men in the launch were not amused, and several times they cut the rope only to have the islanders re-attach it and try again. The crew fired their unloaded arquebuses to frighten the natives off, but then called to the men in the canoes and asked them not to be afraid. That is when the Olosegans became interested in the coiled ropes, arquebuses and swords. One islander tried to grab an unsheathed sword being held by a crewmember and cut his hand open. The canoes then retreated to the beach, but not before leaving with some of the ropes from the ship. Apparently there were gestures and rude words exchanged between the parties that suggested to the Spanish that they were being insulted, and in what might be the future equivalent of road rage, the men in the ship's boats began shooting- this time with ball-loaded weapons. Several islanders were killed or wounded in retreat, and the crew took the opportunity to pick up the severed anchor lines and cables before returning to their vessels. Curiously, some minor details, including the shooting of the islanders and events that followed, are absent from de Quirós' journal, but the descriptions made by Father Friar Juan de Torquemada [13] clarify the sequence of events.

The captain was determined to obtain wood and water that by this time they desperately needed, and to capture at least four island boys to bring aboard. Capture was often practiced for several reasons. Sometimes it was to teach the captives Spanish to better communicate. Other times it was perversely enough, to show that no harm would be done to them. On other occasions, capture was used to barter for supplies, or to convert the taken to Christianity. The intended purpose of this particular kidnap plan was unclear, but another launch of armed men was made.

A contingent of about 150 island men had gathered on the beach in formation and began to shout. It was a war chant and that was followed by a charge, but the Spanish fired their arquebuses at close range killing several. The rest fled in canoes to the opposite shore of the lagoon. The shore party then went inland to the village and began barging into homes. This apparently offended at least one individual who defended his residence by deftly wielding a heavy club. With one of the crew knocked senseless, another of them shot and wounded the defender, but he was undeterred and went after the shooter who then ran him through with his sword. The soldiers then divided into squadrons and marched into the interior where they managed to kill several more islanders, including one who was able to fend off twenty Spanish swordsmen single-handedly, and for quite some time, armed only with a sharpened stick. During this shore mission, the crew discovered two boys, three girls and a young lady, but for reasons unclear, they failed to take them captive as planned. De Quirós thought this to be a rich capture that they left behind, and of course, the pitiable loss of six souls. No water to speak of was discovered, and the natives who were interrogated refused to tell them, or pretended not to understand [13]. However, cultivated food was appropriated for the voyage, along with coconuts and woven mats that were obtained by barter. De Quirós' men rejoiced in the spoils they had obtained, and that none of them had been killed onshore or drowned at sea. The islanders' view of the encounter remains unknown, but was likely to be substantially less joyful.

From Olosega Atoll, the voyage proceeded west, arriving at the Santa Cruz Islands, a group that lies between the Solomon Islands to the west and the present-day island nation of Vanuatu to the south (Fig. 3.5). These are relatively large and high volcanic islands, and completely different in character from coral atolls. There were often much larger native populations to be encountered and considerably more resources to exploit. De Quirós discovered Espiritu Santo, the largest island in Vanuatu in May 1606 thinking it may be part of Terra Australis. His intention was to establish a colony that he named New Jerusalem. Indeed, de Quirós was so imbued with the spirit that he inducted his men into what he called the Knights of the Holy Ghost, whose duty it was to spread God's word here. But during the 35 days they spent in port, chronic food shortages precipitated raids on local villages and produced less than cordial relations between the Spanish and the natives.

Other problems including navigational disagreements between de Quirós and his pilots, bad weather that separated the vessels while exploring to the south, and de Quirós' chronic ill health plagued the cruise [14]. There was also the threat of mutiny. Finally it was agreed that de Quirós should separate from his companion vessel and return to Mexico, where he arrived in November 1606. His voyage may not have resulted in the discovery of Terra Australis, the establishment of a colony, or the conversion of many native peoples, but he was able to chart thirteen atolls and land on two of them, with descriptions of their people and customs from his perspective. His voyage also charted several of the large islands of Vanuatu, and the Santa Cruz Islands, in sum, not a bad record. But despite at least 50 pertinacious requests or "Memorials" to King Philip III begging him to fund a new voyage, de Quirós would never set sail for the Spanish crown again [15].

Text Box: On Latitude and Chronometers

The mariner's astrolabe was an astronomical computer, primitive versions of which had been developed by the ancient Greeks. However, it was the Portuguese who perfected the instrument. Magellan's 1519 voyage carried six of them made of brass, which had all but replaced the earlier wooden versions [16]. Using a rotating sight on the side, pinholes could be lined up with the North Star (Polaris) or the Sun, while the centerline of the device was lined up with the horizon. The angle formed could then be read from 360° lines marked on the perimeter, and the distance north or south of the equator and from the North Pole could be determined by using mariner's tables. There were certain problems with this device. As it was hand-held it was difficult to keep it steady in anything other than calm seas, and an incorrect angle could result in large position errors; five degrees one way or the other would place a vessel nearly 500 km from where it actually was. And measuring the angle of the Sun at noon required looking at it, which is never a good idea, even through a pin-hole. The angle formed by Polaris and the horizon is easier on the eyes than the Sun, but the North Star disappears below the equator, making it difficult to be certain of the ship's position in the southern hemisphere. It would be more than 80 years after Magellan before western navigators were able to correctly map the Southern Cross constellation and use it for navigation below the equator. Eventually, 58 stars, four planets and the Moon could be used to pin-point position by obtaining multiple intersecting lines, but by then the astro-labe would be replaced by a more accurate device, the sextant. Even so, the intersecting lines could only be used to establish multiple intersecting circles on the globe, rather than a single point.

Although maps and nautical position fixing continued to improve in the seventeenth century, they were always off the mark for several reasons. Storms, wind, and sea drift always produced uncertainty in the determination of latitude. Thus blown into the Tuamotu Archipelago in 1606, de Quirós had some difficulty employing his navigational skills to determine which island he called Conversion de San Pablo. In addition to being unsure of his latitude, de Quirós had no way to determine longitude, the imaginary north-south lines connecting the poles that cross the lines of latitude. On an x and y coordinate system, this meant estimating x and guessing at y.

In order to know the position of latitude, the precise time difference between the vessel and a fixed point of land had to be known. The British established such a point at Greenwich England, which defined the Prime Meridian of 0° Longitude. The Distance east or west of the Prime Meridian can be calculated by knowing the time difference between the local time at noon and the exact time in Greenwich (Greenwich Mean Time or GMT). A skilled navigator could use his sextant to track the rise of the Sun and deter-mine the moment that it reaches its highest point in the sky—local apparent noon. Knowing that the Sun moves west at a rate of 15° each hour (at the equator) a local noontime of one hour after noon at GMT means that the longitudinal position is 15° west of Greenwich. At least, that is the theory. The main difficulty until the early 1700s was that clocks were based on a pendulum. As a result their accuracy was compromised on a sea that was even slightly roiled, and a gentle breeze could be enough to throw the pendulum out of its periodic swing. The British sought to solve this problem by establishing a competition through the Longitude Act of 1714 and a prize of £20,000 to develop a seaworthy and accurate timekeeper.

The winner was John Harrison, a carpenter and clockmaker, who without a formal education produced four successively better spring-driven marine chronometers, the last of which was compact, similar to a pocket watch (Fig. 3.6), and contained a lever to maintain even pressure on the mainspring. It even had a special balance designed to compensate for humidity and temperature variations. Unfortunately, Parliament changed the rules in 1774, requiring that the winning chronometer must be tested for a year in Greenwich and further tested on approved voyages. Harrison was awarded only about half the prize and regrettably he died two years later while attempting to collect the rest. Nonetheless the marine chronometer was a major leap forward for oceanic navigation and contributed immensely to explorers from James Cook onward (Fig. 3.7), and their ability to fix and map their positions on the globe.

Captain James Cook   Cook is widely acknowledged as one of the greatest surveyors and navigators of his age. Indeed, Cook's maps of the Great Barrier Reef and the New Zealand coastline, never before mapped by Europeans, were off only modestly compared with those based on modern global positioning satellites. He explored and mapped the islands of Tonga, which became known in the west as the Friendly Islands because of the congenial reception accorded to Cook on his first visit in 1773. Cook also landed at the western Pacific islands of the New Hebrides (Vanuatu) where he was met by hostility, and he was the first European to visit the Island of New Caledonia, as well as the Hawaiian Islands where he met his untimely demise. Cook also landed on Easter Island in the far eastern portion of the Pacific (Fig. 2.7) and spent considerable time in Tahiti documenting the native people there as described further in Chap. 4. Cook did not spend much time on atolls (which he referred to as 'half-drowned islands'), but he mapped the southern Tuamotu Islands in 1767 and 1769 in parallel with the route taken by de Quirós. Indeed, Cook spent more time mapping atolls than he did visiting them. He did visit a few including Palmerston Atoll in what subsequently became known as the Cook Islands. It would be a fitting tribute had it been given by the Royal Navy, but the name in honor of the great English navigator was given when it appeared on Russian naval charts in the early 1800s. Thinking that the name was a good idea, the English eventually adopted it. Cook also visited the elevated atoll called Christmas Island on Christmas Eve 1768, which was commemorated by the translation of 'Christmas' into the local language as 'Kiritimati'. There is no question that the good captain knew where he was.

Fig. 3.6 Harrison's compact marine chronometer displayed at the National Maritime Museum, Greenwich (Wikimedia Commons)

Fig. 3.7 Oil on canvas portrait of Captain James Cook, 1776, National Maritime Museum (Courtesy of Wikipedia. Org)

References

1. Kodicek EH, Young FG (1969) Captain Cook and Scurvy. R Soc J Hist Sci 24:43–63
2. de Silva CR (2009) Portuguese encounters with Sri Lanka and the Maldives. Translated texts form the age of discoveries. Ashgate Publishing, Burlington
3. Dunn RE (2012) The adventures of Ibn Battuta: a Muslim traveler of the fourteenth century. University of California Press, Berkeley
4. Al Suood H (2014) The Maldivian legal system. Maldives Law Institute, Malé
5. Hogendorn J, Johnson M (1986) The shell money of the slave trade. Cambridge University Press, Cambridge
6. Thomas H (1997) The slave trade. Simon and Schuster, New York
7. Robertson JA (1906) Magellan's voyage around the world volume 1 by Antonio Pigafetta. AH Clark Company, Cleveland
8. Fitzpatrick SM, Callaghan R (2008) Magellan's crossing of the Pacific using computer simulations. J Pac Hist 43:145–165
9. Stanley, Lord of Adderley (1874) The first voyage round the world by Magellan, translated from the accounts of Pigafetta and other contemporary writers. The Hakluyt Society, London
10. Hale EE (1890) Magellan and the Pacific. Harpers New Mon Mag 81:357–366
11. Rogers RF (1995) Destiny's landfall, a history of Guam. University of Hawaii Press, Honolulu
12. Markham C (1904) The voyages of Pedro Fernandez de Quiros. Translated from the original Spanish manuscripts, vol. 1. The Hakluyt Society, London
13. Markham C (1904) The voyages of Pedro Fernandez de Quiros. Translated from the original Spanish manuscripts, vol. 2. The Hakluyt Society, London
14. Freeman DB (2010) The Pacific. Routledge, London
15. FMH D (1961) A catalogue of Memorials by Pedro Fernandez de Quiros, 1607–1615. Trustees of the Public Library of New South Wales, Sydney
16. Mathew KM (1988) History of Portuguese navigation in India, 1497–1600. Mittal Publications, Delhi

Chapter 4
Import, Barter and Trade, and the Natural Resources of the Pacific Islands

Abstract This chapter reviews the impact of non-native species, including those brought first by the Lapita people of the Pacific as they colonized islands and moved breadfruit taro, pigs, dogs, chickens and Pacific rats with them. The ecological effects of those imports were later overshadowed by exchanges with European and American visitors that permanently altered the island landscape. The Tahitian exchange is the best known of these interactions, and was emblematic of the broad and lasting impact on native life, land and culture, with each side working for its own purposes. The importation of European domestic animals and plants, as well as the accidental introduction of new rats and microbes was a prelude for later importation of alien snakes, and other invasive species. In addition, trade between east and west altered the economy of the tropical Pacific islands, extirpating or diminishing natural resources ranging from sandalwood forests on land to whales at sea.

4.1 Domestic Animals and Plants

One of the hallmarks of the Lapita people who first colonized the islands of Remote Oceania 3000–3500 years ago was the transport of starchy root crops including carefully wrapped taro tubers and breadfruit seedlings, as well as live chickens at first, then pigs and dogs later. Rats, perhaps incidentally or on purpose also accompanied the early voyagers to new shores [1]. Polynesians discovered Tahiti and the surrounding Society Islands nearly 2000 years before the Europeans and they had developed considerable skill in managing their natural resources. That is not to say that they had little impact on the land. They kept pigs that uprooted forest floors, converted marshlands into taro fields, and cut terraces into hillsides to plant trees that they had brought with them [2]. However, even with those skills and the societal elements that reinforced them, heavy rainfall or droughts as well as unpredictable tropical cyclones brought food shortages and even famine to the region. The native islanders would later also have to adapt to the demands of provisioning European ships. The natural resources that were available on tropical islands of the Pacific and their role in trade are outlined in the following sections.

© Springer International Publishing AG 2018
W.M. Goldberg, *The Geography, Nature and History of the Tropical Pacific and its Islands*, World Regional Geography Book Series,
https://doi.org/10.1007/978-3-319-69532-7_4

Chickens

Chickens that were brought along with the first Pacific island colonists were descendants of the Southeast Asian jungle fowl *Gallus gallus*. They were raised mainly as food and for their feathers, but curiously, the eggs were not eaten, and as is the case with other valuables they were also offered as religious sacrifices. Chickens were cared for and hen houses were found around the sites of the most ancient villages or houses [3]. The original Southeast Asian chickens (Fig. 4.1) are associated with a unique genetic marker that can be used to trace their migration, the route of which is somewhat controversial. Studies suggest that these animals were transported from the coast of New Guinea and the Bismarck Islands to Micronesia about 3850 years ago, without further onward transportation into Polynesia. By contrast, chickens currently found in Polynesia appear to have originated later through Vanuatu and other southern islands, then moved farther eastward from there. Authentic Polynesian chickens appear to represent an admixture of the ancient Pacific island group and subsequent evolution within Polynesia. However, both of those populations are genetically distinct from chickens found in South America [4]. This contradicts the evidence suggesting that Polynesians crossed the Pacific and brought South American chickens back with them, unlike the route of the sweet potato described below.

Dogs

The history of dog domestication is often depicted as a two-stage process where primitive dogs were first domesticated from their wild ancestors, the gray wolves, after which the early dogs were further selected to form many breeds with

Fig. 4.1 The jungle fowl *Gallus gallus* native to Southeast Asia, spread throughout the Pacific islands after their early transport to New Guinea and Far Oceania (Courtesy of Wikipedia.Org)

specialized abilities and morphology. However, despite many efforts focused on the study of dog evolution, several basic aspects concerning their origin and evolution are still in dispute. It is not clear which type of dog accompanied the first settlers of the Pacific islands, although there are genetic data that suggest the original Polynesian dogs are related to Australian dingoes, which appear to have spread from mainland Southeast Asia to Indonesia and Melanesia, then to Oceania, likely more than 8000 years ago based on genetic information [5, 6]. In any case, there is indisputable evidence of dogs in Polynesia by 2000 years ago, taken by the original island settlers along with other domestic animals [4]. A short-haired breed that rarely barked was used as food in Tahiti, although until that occasion arose they were also considered as pets. A long-haired breed, possibly a spaniel, reached the Tuamotu Islands in 1526 by a Spanish vessel, and by the time of Captain Cook's visit more than 200 years later, both may have been present in Tahiti. They were often considered as good as pigs to eat and both were consumed on special, ceremonial occasions [2, 3].

Pigs

Pigs in Polynesia and elsewhere in the Pacific were *Sus scrofa*, which had been domesticated from wild boars in Asia for thousands of years, but arrived in peninsular Southeast Asia, New Guinea and islands of eastern Indonesia 3500 years ago. These pigs have a Pacific genetic signature that can be traced throughout Melanesia and Polynesia, and their dispersal is closely associated with the Lapita and later Polynesian migrations. There is also a genetically identifiable East Asian group of *S. scrofa* that was likely introduced from mainland Asia and Taiwan into the Philippines, the Mariana Islands and into parts of western Micronesia [7, 8]. Five other species of 'wild pigs' are found in various parts of the Philippines and Indonesia that are distinguishable from *S. scrofa* on the basis of distinct genetics and dentition [9].

Male pigs (boars) or the intersex variety grow tusks (Fig. 4.2) and these have long been venerated in traditional Vanuatu societies (formerly the New Hebrides Islands). Well-endowed tuskers were capital. They were used as currency, determined social standing and credit worthiness. Indeed, the entire economy was (and in some places still is) predicated on the number of such pigs that a man possessed. They are still used as payment for land, services, sacrificial feasts, and are traded for other valuable goods. They may be used for bride money, blood money, ransom or payment of fines for committing acts that are taboo. An influential member of the community called a *bigman* held that status as a function of his tusked possessions [10]. Chiefs, of course, had more of them than others, but both bigmen and chiefs still wear large pig tusks as symbols of rank and authority. Pig loans often rely on tuskers as credit instruments on some islands. A creditor could allow a pig to be borrowed with interest based on how much the tusks had grown during the term of the loan. A pig with a half circle tusk would be worth four times as much if returned

Fig. 4.2 Pig with a half-circle tusk from Pentecost Island, Vanuatu. Image from Flickr.com courtesy of the photographer

years later with a full circle, and ten times as much with a circle and a half. Most of these transactions took place in a public marketplace like a stock exchange, or a bourse as they are called in Europe, and thus the entire porcine enterprise on Vanuatu has been referred to as a 'pig bourse' [11].

Pigs likely arrived in the Society Islands about 1700 or 1800 years ago and were an integral part of Tahitian society by the time of first European contact. They were described as small, 40–60 pounds with relatively long legs, similar to those in China, but "plump and well flavored". Europeans brought pigs to Tahiti in 1774. They were the same species but were much larger, stockier and paler than the Tahitian variety. Within a few years Captain William Bligh would report that the islanders preferred the chubbier pigs and it was not much later when the standard Tahitian pig had doubled in size [2]. As in Melanesia, pigs were standard fare for wedding payment, gifts and trade, and their possession enhanced the social prestige of the owner. Although pigs were eaten by the upper classes they were first sacrificed as a means by which spirits could answer questions through a priest. The animal was typically strangled to avoid losing blood (considered a delicacy). The pig was then disemboweled, the blood drained and its belly fat removed. The body cavity was then filled with hot stones and placed to cook in a stone-lined ground oven. A priest would seek to interpret what a spirit sought to communicate from observing the animal's death struggle. Whether the eyes of the dead animal were equally open, or if the mouth was closed or showed teeth signified different omens, especially regarding warfare. The intestines were examined immediately after death and were another source of information concerning future events [2].

Breadfruit

Breadfruit has been an important natural resource in the Pacific for more than 3000 years. Three close relatives are recognized, the most ancestral of which is the breadnut, native to New Guinea and Indonesia, the fruit of which is diminished compared with its large edible nuts. Breadfruit per se, *Artocarpus altilis*, is about the size of a cantaloupe or a large grapefruit and is recognized by its hard and pocked-marked skin (Fig. 4.3). This is a domesticated species that is a hybrid primarily between breadnut and breadfruit and it is typically sterile, lacking seeds [12]. A third species grows in the Caroline and Mariana Islands and does especially well on atolls, as long as there is a freshwater layer that can be accessed by the root system. The original progenitors of breadfruit were brought wrapped as cuttings or saplings east from Near Oceania but the species is now widespread and has been transported by ship throughout the tropical world.

Breadfruit had been naturalized in Tahiti and other Society Islands long before European contact. Bland, starchy and inedible unless it was first roasted, boiled, steamed or fermented, breadfruit was nonetheless a staple and a revered food source. Although the tree was easy to care for, could grow more than 18 m and yield 150 fruit per year, it was treated with great care. Groves were planted on coastal lowlands, and the islanders could name fifty varieties. Some of these could be planted on high island hillsides at various altitudes to take advantage of later fruiting. This characteristic was especially useful as the dry season approached (June-December), a time when most plants did not fruit. The landowning class typically had at least one breadfruit tree on their properties, although groves often belonged to families of high standing or to groups of families. Ripe breadfruit would typically be placed in a stone- and wood-lined pit that was set alight. After 3–4 days the oven was opened, the blackened skin peeled away and the white flesh now brown from cooking was eaten.

Breadfruit also could be fermented by placing it in a pile above ground for several days, followed by burial in pits covered with stones and earth. Months and even years later, the gooey fruit could be eaten and it was often used as famine food. Starvation conditions included prolonged droughts, heavy El Niño rainfall, and warfare. In the latter case, destruction of the enemies' food sources, even hidden fermenting breadfruit, was a typical practice. However, there was more to breadfruit than the fruit. The trees provided abundant shade. The wood was used to make canoes and religious carvings. The sap was used as glue, the leaves, bark and parts of the fruit were applied as a poultice for wounds, and the leaves and inedible parts of the fruit were used to feed pigs and dogs [13].

Taro and Other Root Vegetables

Among the many introduced plant species that originated in tropical Asia, root crops were significant and included two distinctive large-leaved 'elephant ear' plants, taro *Colocasia esculenta* and giant swamp taro *Cyrtosperma merkusii*. Taro

Fig. 4.3 Ripening breadfruit (Image ©Jim Wiseman, courtesy of the Breadfruit Institute, National Tropical Botanical Garden, Honolulu)

is an herbaceous, lowland plant that grows to a height of 1–2 m. These plants do well in deep, moist or even swampy soils where the annual rainfall exceeds 2500 mm. The edible part of the plant is typically a swollen, underground plant stem covered with leaf remnants called a corm that serves as a starchy and hardy storage organ (Fig. 4.4). Older corms often develop multiple 'cormlets' that can be dug up, separated and re-planted, each giving rise to new plants. Leaves can be cooked to produce a spinach-like product. Unfortunately, taro and other elephant ear plants are rich in needle-like crystals of oxalic acid that cause irritation of the mouth and throat unless they are soaked in water before being boiled, steamed or baked.

Fig. 4.4 Taro, *Colocasia esculenta* showing the starchy root (*left*) and the 'elephant ear' leaves (*right*) (Images courtesy of Wikipedia.org)

Various lines of evidence suggest that it originated in south central Asia, and was then spread to Southeast Asia. Natural populations of taro may also occur in the Bismarck Islands, but its movement through Far Oceania corresponds to Lapita settlements in Vanuatu and Fiji [14]. From there it has long since expanded throughout Polynesia where it is intensively cultivated and contributes its highest percentage of calories to the diet. It can be prepared much like potatoes, eaten whole, mashed into a paste that Hawaiians call poi, or cut into thin slices as chips.

Taro is quite variable in form and in tolerance of conditions, but it is typically grown in unflooded areas of high rainfall (upland taro), although lowland (wetland) taro can be grown in waterlogged or flooded fields. The upland form is ready to harvest after 8–10 months when the leaves begin to turn yellow and wither. Wetland cultivation requires specialized conditions including marshy areas of freshwater springs, stream banks and most commonly in artificially irrigated pond fields. The latter is labor intensive, and requires constant maintenance and care. The crop grows from corms placed into the muddy soil but the water level must be gradually raised to keep the base of the plant submerged. In addition, water flow is necessary at all times to prevent low oxygen from developing in the soil and to dissuade attack by organisms that promote rot [15].

Fig. 4.5 Swamp taro under cultivation on Butaritari Atoll, Kiribati (Image courtesy of Wikipedia.Org)

Swamp Taro

Giant swamp taro (*Cyrtosperma merkusii*) is an alternative crop for areas that are wet, but are without flowing water. It is similar to true taro, but this species is in a different genus. It has larger leaves, grows up to 4–5 m tall and has much larger, coarser corms that can weigh more than 80 kg (Fig. 4.5). Because of its size it has to be cooked for longer periods than true taro. It can be 'field stored' (left in the ground without harvesting) for very long periods – up to 30 years or more – and accordingly has traditionally been an important emergency crop in times of natural disaster and food scarcity [16]. Swamp taro is also an important source of carbohydrates on atolls and it is an important supplement to a diet that may otherwise consist of seafood and coconut. Like true taro, it was cultivated throughout Southeast Asia and was likely brought to the Pacific by the Lapita. Its primary advantage is that it can be cultivated on harsh coral atolls such as those in Tuvalu and Kiribati (see frontmatter map Fig. 1.0) in the central Pacific where it is grown in hand-dug pits that contact freshwater. Some of these excavations are 90 m long by 60 m wide and are nearly 2000 years old [17].

This layer of fresh water is typically recharged from rain and may occur as a thin layer that floats directly on the top of denser seawater. Swamp taro can even tolerate a small amount of dilute salt water (i.e., brackish water), but it does not do well in barren limestone soil unless the pits are enriched with humus from decaying leaves. Sometimes such material has to be imported by canoe, but animal fertilizer was apparently never used in Polynesia [16]. On coral atolls and other similar low islands, the highest elevation is often the spoil banks from pits excavated for swamp taro [17].

Yams and Arrowroot

There are several species of yam, vines with starchy roots that are sometimes confused with taro. Yams, like potatoes, grow as stem tubers rather than from corms. These are short underground stems capable of forming shoots that grow into typical stems, leaves, and roots. Depending on the species, stem tubers can be small and round or they may produce a longer cylindrical shape. They typically have a black or brown bark-like skin and white, purple or reddish flesh. Yams were likely introduced to the Pacific islands through Southeast Asia along with several other root vegetables, including arrowroot, *Tacca leontopetaloides*. This distinctive member of the yam family is native to Southeast Asia and New Guinea, and was taken with early settlers of Remote Oceania, Polynesia and Micronesia [18]. Polynesian arrowroot forms from an underground stem that contains a different type of starch. It is first grated and then allowed to soak in fresh water. Upon settling, the starch is rinsed repeatedly to remove the bitterness and then dried as a powder. It has no flavor of its own, so it can be used as a thickener like cornstarch. In Polynesia it is mixed with mashed taro, breadfruit or sweet fruit extract, or with coconut cream to prepare puddings. Arrowroot is an important food source for many Pacific island cultures, especially for the inhabitants of low islands and atolls.

Sweet Potato

The sweet potato, *Ipomoea batatas*, is a root vegetable that forms differently from the stem-tuberous yam. Instead, the sweet potato is a root tuber that is used in some parts of Polynesia, particularly in the highlands of New Zealand and New Guinea. It has the advantage of storability in specially designed warehouses and although it is not an historic staple of most Pacific islands, it is now widely cultivated where the land is sufficiently fertile [19]. It also has an interesting and complex history. An examination of modern and ancient sweet potato genetic lineages now suggests a three-way transport to Polynesia. The first appears to be a prehistoric transfer from the area of Peru and Ecuador about 1000 years ago by Polynesians who crossed the Pacific and brought the sweet potato back with them. Polynesian sweet potatoes were part of a distinct lineage that was native to South America and predates the introduction of different genetic lineages introduced more than 500 years later by early European explorers. However, such an exchange appears to have taken place as a second introduction of the sweet potato to Polynesia. A more complex combination occurred later when the European and South American varieties were hybridized and spread again to Polynesia for the third time [19].

Coconut Palms

The coconut palm (*Cocos nucifera*) likely originated from Southeast Asia [20], but is now found throughout the tropics where it has been widely cultivated on plantations. Indeed, they became so widespread that the coconut has became characteristic of the south Pacific even though in many places such as the Tuamotu Archipelago (French Polynesia- see Fig. 5.4) they were thought to be rare or non-existent before the 1840s when commercial plantations were initially developed [21]. Cultivation of this tree is a testament to its great versatility as evidenced by the diets of many people in the tropics and subtropics, and the widely tolerant conditions under which the trees will grow. The immature nut contains a clear liquid (coconut water) that is rich in carbohydrate and electrolytes that can serve as a lifesaving drink in areas where there is no other source. In time that liquid turns into a layer of hard, white flesh on the inside of the shell. Softer embryonic material forms later, just before the nut falls from the parent tree for dispersal by waves and currents (Fig. 4.6). The nut with its self-contained supply of nutrients can float and remain viable for up to 110 days [22]. The flesh can be scraped and dried as copra, or it can be pressed to extract coconut oil or cream. The oil is initially yellow and is clear after processing, and can be used in cooking or as a skin lubricant.

In Europe, coconut oil was made into soap with the addition of glycerin and lye, or was used as the base in the manufacture of cosmetics. A brief search on the Internet will reveal that it is also sometimes touted as a cure for nearly every human malady known, similar to the way snake oil was marketed a century ago. Other parts of the tree are also useful. The trunk can be used for charcoal, lumber, or drums and other musical instruments. Thatched roofs can be fashioned from the fronds, caulk can be made from the sap, and palm sugar can be produced from the flower nectar [23]. The fibrous husk of the mature nut, called coir (see Chap. 3), can be woven into mats, string, fishing nets or rot-resistant rope. The hard stone-like layer surrounding the mature flesh can be made into bowls, utensils and musical instruments.

Coconut palms do not due well in cooler climates, although they may grow without producing coconuts at the margins of the tropics. Under more favorable conditions the tree will fruit within five years of germination and will continue production for 40–60 years. They prefer sandy soil and require warm sunny climates with regular rainfall and high humidity. However, on atolls the trees must tap into a freshwater lens to survive and they are very sensitive to water shortages [22]. The droughts associated with El Niños therefore caused many copra plantations to collapse during the late nineteenth century. Those found on remote atolls, for example the ones on central Pacific islands, were quickly abandoned for that reason. Even larger operations faced problems. In the early 1900s The Lever Brothers formed an English company (forerunners of the multibillion dollar Unilever Corporation), and with the idea that coconuts were the future, purchased thousands of acres on the Solomon Islands for plantations. This scale was far greater than could be expected from atolls, but the price for coconut products could not sustain their investment, even while using indentured labor. However, it was the invasion of the Solomon Islands by Japan rather than market forces that ultimately forced the Lever brothers to abandon their plantations [23].

Fig. 4.6 Typical coconut palm and coconut sections. (1) section through mature fruit showing fibrous outer husk and hard-shelled kernel with white flesh; (2–3) stony inner portion of coconut and cross-section of the kernel; (4) stony outer covering removed to show the outer side of the flesh; (5) embryo forming within developed coconut (Image courtesy of Wikipedia.org after Köhler 1897)

Pandanus

This highly diverse group of more than 740 species [24] is recognized by its prominent stilt-like root system. They also form palm-like fronds that develop a swirled or screw like pattern (Fig. 4.7). That characteristic gives it the common name of screwpine, although they have nothing in common with pine trees. Likewise, while they are often called pandanus palms, they are not closely related to palm trees either. Such is the problem with common names. They vary in size from small

Fig. 4.7 *Pandanus tectorius*, the screwpine, with its typical stilt-like roots and whorled palm-like fronds (*top*). The unripe green fruit (*bottom*) shows its characteristic wedge-shaped 'keys' that will turn yellow to red-orange when mature (Images courtesy of Wikipedia.Org)

shrubs less than 1 m tall, to medium-sized trees that reach to 20 m, and typically have a broad canopy that is heavy with fruit. Although some species are capable of growth at altitudes of more than 3000 m, the best-known is *Pandanus tectorius*, which is well adapted to coastal lowlands, and in terms of cultural and economic importance in the Pacific, they are second only to coconut palms on atolls [25]. This pandanus can withstand drought, strong winds and salt spray. They propagate readily from seed, and like breadfruit they can grow from branch cuttings. This species does not do well in dry environments and requires at least about 1500 mm of rainfall per year. Like coconut palm fronds, the leaves can be sliced into fine strips and woven into mats, ropes, baskets, canoe sails, thatched roofs, and grass skirts. Its exact origin is not known, although the original voyagers may have introduced *P. tectorius* to the tropical Pacific from Indonesia and SE Asia. Its fruit varies in size and shape but is generally round and is composed of up to 200 tightly bunched, wedge-shaped structures called keys (Fig. 4.7) that are yellow to dark red-orange when ripe. The flavor has the sweetness of coconut with a hint of citrus and a nutlike flavor from the seeds in some varieties, although other species that produce relatively small fruit can be bitter or astringent. The seeds within the fruit can float and are viable after being at sea for many months [25].

Bananas

These familiar fruits are all members of the genus *Musa* and are native to the Indo-Malaysian region, Southeast Asia and New Guinea. The latter locale is home to the greatest variety of types including round fruit, and differences in skin and color (yellow, green, red, silver). Different groups can be starchy (e.g., plantains) and require cooking, whereas other varieties are sweet and are eaten raw. The earliest indigenous voyagers carried bananas throughout the Pacific where they have become naturalized. They are the source of food and fermentable sugars used to make beer, wine and vinegar. Stalks were traditionally used as medicine including poultices for sprains, peels as bandages. Root sap treated warts and could be used as a black dye. Thread, rope and skirts could be made from the strong fibers of the outer stem and from leaf fibers. Leaves could be also be used for thatched roofs and wall lining, as a wrapper for cooking, and as tobacco paper for smoking. The shape of the fruit is sexually suggestive and it was held in high esteem in some island cultures where women were not allowed to touch or eat them. The plants arise from an underground, starchy rhizome that was often used as famine food when the necessity arose. Bananas generally grow well on high islands and on atolls, especially in areas where rainfall is evenly distributed throughout the year. Dwarf varieties can tolerate longer dry seasons [26].

4.2 Marine Resources

Fishing Techniques [2, 27]

There is a considerable difference between high and low islands in natural resource availability. Atolls in particular are often without fresh water as the early Portuguese and Spanish explorers quickly learned (Chap. 3). Inhabited atolls were limited to those with access to freshwater, and while lagoon fishing provided protein that could be supplemented by coconuts, swamp taro and pandanus fruit, such islands were composed of limestone soil that typically lacked the ability to support domesticated animals and plants. However, they were very well stocked with marine life.

In Pre-European times, the Polynesian societies had a unique connection with the sea. They not only saw it as a source of food, but because all marine animals were descendants of Tangaroa, god of the sea, they represented their gods as well. Polynesians were skilled fishermen, and they had names for at least 150 different species. Techniques for catching fish were quite sophisticated, taking into account the phases of the moon and the condition of the tides during which certain types would be on the move. Hooks were made from bone, mollusk or turtle shell, or volcanic stone, typically with a strong inward curve and perhaps a barb, depending on the type of fish to be caught. Baited hooks could be used with rope made from coconut fiber in shallow water. Alternatively they used bamboo poles along with shiny shell lures deployed from moving canoes to target offshore fishes including mackerel, dolphin and tuna. Reef fishes, squid and octopus were also chased underwater and speared as fishermen held their breath for 2 min or more.

Netting made of coconut fiber and poles were constructed to span the channels between islands, distances that could be hundreds of meters and sometimes required the participation of an entire village. Larger fishes were chased toward the net, which was weighted at the bottom by stones and gradually closed at the ends by forming a circle that continually diminished in circumference. Stone fishing was a variation of this technique that employed a cattle-drive like method of herding fish into nets with noise made from clapping stones together. Driving them into shallow water, a long line of coconut leaves was dropped around them and from within this trap they were harvested using woven baskets as a strainer. Other fishing methods employed box-like traps that were hidden on the reef and baited with crabs or fish heads. The collection of marine animals and seaweeds at low tides from the reef flat and lagoons was also an important source of food, the work typically accomplished by women and children. Species taken in this practice of reef gleaning included sea urchins, crabs, shellfish, eels, octopus and finfish trapped in tide pools.

Naturally occurring fish poisons have been used in Polynesia, but they were employed more often in Melanesian islands where plant diversity is higher and the choices are wider. In some species, the roots are used whereas in others, the toxins reside in the stems, leaves or nuts and are readily extracted by pounding, grating or boiling them. These poisons stun fish on exposure, and although the chemical manifestations of the toxins differ, they typically interfere with oxygen transport across the gills and are not toxic to the fishers. In order to be effective, such agents had to be used in enclosed basins such as those forming on reefs at low tide or in streams or ponds.

4.3   The Tahitian Exchange

Exchange is a powerful social operation that encompasses gift giving, trade and barter, especially between previously isolated cultural groups. The first great exchange occurred between Europe and the discovery of America in the wake of Columbus' expedition in 1492. This so called Columbian Exchange occurred on a massive scale and included export from South and Central America of unusual plants including vanilla, pumpkins, pineapples and corn among many other once-regional items that became global in a relatively short period. Potatoes found their way to Ireland, maize and tomatoes to China. In return various grains, fruits and livestock native to Europe were transported to the Americas, as were technological advances in manufactured goods including cotton and wool clothing, metalwork, naval architecture, and of course weaponry. Multiple diseases and slavery were also part of the Columbian Exchange, although those were typically unidirectional 'gifts' from the dominant culture. At the same time, similar exchanges were taking place in coastal Africa, India and Indonesia as described in Chap. 3.

A far smaller exchange took place in the eighteenth century when Europeans encountered Pacific islanders who until that time existed as Neolithic native cultures. Some were less inclined to embrace the first explorers, but others, especially the Tahitians and other indigenous Society islanders found that the Europeans could be used for the their own gain. Some perspective of what came to be known as the Tahitian Exchange [2] can be gained from the enormity of undertaking the long voyage from Europe. Such ventures typically lasted 2–3 years. Ships were small, and crowded with equipment and stores including salted beef and pork, butter, oatmeal, casks of soup, water, and rum. Bread was taken aboard in sacks. However provisions often did not last. Rats and weevils attacked the bread. Live animals including chickens for eggs and meat, and goats for milk were brought along, but like the sailors, the animals suffered from various ailments and storms that would carry some of them off the ship.

The European Introduction

Captain Samuel Wallis and his 24-gun frigate HMS Dolphin left England in August 1766 taking five months to reach the Straits of Magellan at the tip of South America, and it was not until June 1767 that land was sighted. The captain and crew did not know if I what they saw was an island or the tip of the great and elusive continent of

Terra Australis, but after 10 months at sea, the captain and his men were ill, a combination of sundry maladies including seasickness, scurvy and diseases induced by near starvation. Whatever they saw in the distance made them hopeful and pleased, but it was not entirely clear that the Tahitians were happy to see the Dolphin. As the ship attempted to anchor, skirmishes broke out. Canoes loaded with stones were used to pelt the ship, and attempts were made to ram the smaller boats as they sounded for anchorage or tried to go ashore for fresh water. When the ship anchored, a flotilla of about 300 canoes approached, apparently with fruit on banana leaves and naked women "practicing allurements" [2]. But when the signal was given, the Tahitians reached under the leaves and started throwing rocks. The larger canoes were more than 25 m long, each powered by 40 warrior-rowers despite the painting (Fig. 4.8) that depicts the canoes as much smaller than the 34 m-long Dolphin. In addition, the 130-member crew was vastly outnumbered. That produced some nervousness and musket shots rang out, followed by cannon fire (Fig. 4.8) that splintered canoes and killed many native men and women. No body counts were made.

After this incident the ship's officers landed, protected by marines, and by planting the Union Jack claimed the island for king George III. The next morning 'several thousand' Tahitians were seen carrying the flag away and Wallis ordered cannon

Fig. 4.8 The HMS Dolphin under the command of Captain Samuel Wallis, fires upon the natives of Tahiti who greatly outnumbered the English. Note the small size of the canoes depicted, despite some of them being nearly as large as the Dolphin. National Library of Australia, 1767 (Artist Unknown, courtesy of Wikimedia.Org)

fire; it was unknown how many were killed. For Tahitians, these incidents required giving gifts normally reserved for gods, including eight large pigs, four piglets, two dogs, a dozen chickens, fruit and large bails of bark cloth. In return, Wallis left two turkeys, two geese, three guinea hens, a pregnant cat, china, glass objects, needles, thread, peas, beans and seeds of several types of fruit trees [2]. The first of many exchanges had been made and the trade for provisions had begun. Wallis' cruise would be followed by Bougainville, Cook, Bligh and other captains, as well as the missionaries, traders, whalers and other European vessels that anchored in Tahitian waters. Indeed, bulls and cows, donkeys, horses, rabbits and even guinea pigs were brought not simply to trade, but to populate the Society Islands for future resupply needs, and perhaps to be of some benefit to the islanders themselves.

Tahitians found European goods attractive and considered them fair game despite harsh reprisals for what the British perceived as ungentlemanly behavior. Tahitians were shot at, hauled out of the water with boathooks, flogged, and ears were lopped off to dissuade the natives from their five-fingered approach to trade. However, Tahitians soon learned the art of negotiation. Even while Captain Wallis attempted to keep his men in line by flogging, women learned that by using unmistakable hand and finger gestures that they could acquire more valuable and larger iron nails in trade for certain favors. After a few weeks of this trade it was discovered that nails were being taken from everywhere, even the cleats holding the HMS Dolphin's hull together. Reef fishes, taro, chicken and pigs could be traded for nails, scissors, iron axes and European clothing, which soon became haut couture among the natives. Women soon began wearing tablecloths as dresses, white sailor hats, mariner's shirts, and even red coats with embroidered gold buttonholes when they could get them. Tahitians also became 'collectors', that is, they acquired the European practice of simply having objects for their own sake, rather than using them for trade or consumption [2]. It was apparent that the traditional culture was being altered in irreparable ways.

The Salt Pork Trade

Tahitian chiefs had control of natural resources and could declare a formal restriction on them, either for management purposes in times of shortage or to control trade. Captain James Cook noticed on his second voyage to Tahiti in 1773 the Tahitians began hiding their pigs and other comestibles as Cook sailed into port. Higher prices in trade were expected to remove restrictions, often newly imposed by the chiefs on sight of the arriving vessel. In addition, an evolution of trade expectations became evident. In 1767 a large nail could fetch a pig sufficient to feed 20 men, but ten years later Captain Cook noted that an axe was required to close the deal. Nails no longer interested them [2]. This is not to say that Tahitians were engaged in price gouging, but rather, that their food supplies had been diminished by the number of European ships that had stopped in Tahiti for supplies.

The Tahitian economy was beginning to change, and the degree of change was about to take a sharp turn. In 1788 the English had established a penal colony in New South Wales, Australia where 2000 convicts as well as military personnel had to be fed. The food came from England and it was expensive to import, especially the salted pork, which consumed more than 60% of the funds allocated for food. In 1801 the Governor of New South Wales sent a ship to make a trial run to Tahiti and establish a more local source of pork by trading the usual iron tools among other items. Tahitians were required by the regional leader at the time (King Pomare II) to bring their pigs to shore where they were promptly butchered, salted and stored in casks of brine. Several hundred pigs later, 31,000 pounds of pork had been loaded onto the ship bound for Australia. This was the first truly commercial European expedition to Tahiti, and it would soon draw the island into a trans-Pacific, international trade economy. It was not long before small ships from Australia would be bartering for 400–500 pigs, and larger ones operated by private merchants would pack in the meat of 2–3 times that many [2].

Thus began the Tahitian salt pork trade, and during the 25 years of its existence a total of more than 3,000,000 pounds of pork had been traded. That was certainly a lot of pork for the islands to provide, and King Pomare II recognized the forces of supply and demand in establishing exchange values. As shortages developed he imposed taboos on the consumption of pork by commoners. While some of the pigs came from the outer islands, most of them came from Tahiti. Each family may have been pressured to give up as many as nine or ten pigs per year, and breeding pairs were likely lost. Being mindful of his enemies and his ability to wage war against them, Pomare bargained with whomever he could to exchange pigs for firepower.

The government in New South Wales began to 'outsource' their procurement of salted pork and sent a privateer vessel, the Harrington, to Tahiti in 1807 for that purpose. The transactions described by missionary William Ellis were as follows: "The kings and chiefs would receive nothing in exchange for the produce of the islands except fire-arms and powder. Even the wretched females whom the king sent off to the ship for the vilest of purposes were ordered to take nothing as the price of their debasement and vice but gunpowder, which they were strictly enjoined to bring to the king. Such was the eagerness of the people to obtain arms and ammunition" [28]. Such armaments included not only muskets, but also swivel guns and a cannonade, a cannon mounted on a carriage that could fire a solid ball at short range. One has to wonder what was traded for such weaponry.

Nonetheless, Pomare was unable to protect his people from frequent attacks by rival communities, and there was resentment on his demands for the pig trade. There were also diseases being brought to the island by ship, and cultural issues that developed with the missionaries (Chap. 5). It was a tumultuous time and Pomare did not live long enough to see the fruits of his labors. After his death in 1821, the unsustainable Tahitian pork trade began to wane and by 1826 it had ended. It is not clear how the domestic pig population fared in the 1830s, but by 1842 the French took over and promptly regulated and controlled the trade.

The Export of Breadfruit

Interest in breadfruit as an export began shortly after its discovery by the Europeans in Tahiti in 1767. However, by the 1770s England was on several battlefronts with the Spanish, the French, and the American colonies. America had been a food source and was especially critical to sugar cane and other plantations that had been established in the British Caribbean. In Jamaica alone, hundreds of thousands of African slaves had been imported. Following independence, the Americans established a trade embargo that exacerbated already existing food shortages due to hurricanes (five in six years) and droughts [29]. An estimated 15,000 slaves died of starvation or diseases related to it [30]. Thousands of others were in a state of near starvation. An inexpensive, easy to prepare food was sought, and plantation owners began to lobby Parliament, the British Navy, the Royal Society, and anyone else who had influence, to fund a voyage to Tahiti and bring breadfruit to grow in the Caribbean. Sir Joseph Banks was a British naturalist, botanist, advisor to King George III, and a person of means. As the long-time president of the Royal Society he became an influential man of science and a patron of it as well. He had accompanied James Cook on his first voyage to the Society Islands and was very familiar with breadfruit. Mr. Banks took up the cause.

Other plants could have been selected including sweet potato, yams, bananas and various apples, but breadfruit had come to represent a simple, nutritious and wholesome food. To make matters urgent it was rumored that the French were about to import breadfruit to the Caribbean, and that could hardly be countenanced. Mr. Banks soon had the ear of important people and the HMS Bounty, with its soon-to-be infamous William Bligh as captain, set out to Tahiti for breadfruit. As is now well known from the 1932 novel Mutiny on the Bounty, and the movies that followed, Bligh remained in Tahiti for more than five months during 1788–1789 and packed more than 1000 breadfruit plants in pots for the journey to Jamaica. Bligh was a firm disciplinarian whose short temper was unfortunately combined with a confrontational style. He had his men on half rations even while in Tahiti as a means of enforcing discipline as he was preparing to leave. Many of the men were reluctant to do so, having enjoyed life on the island. But after short rationing continued and the crew finding themselves watered less often than the plants, a minority decided to free the breadfruit (in the ocean) and take over the ship. The mutiny took place 24 days after departure, 2000 km from Tahiti. Bligh blamed it on hedonism and poor discipline despite testimony at trial to the contrary. Nonetheless the British Admiralty exonerated him.

Bligh returned to Tahiti for a second breadfruit expedition two years later in another ship, this time accompanied by Royal Marines to preclude another mutiny. On this voyage he was able to secure nearly 2400 breadfruit trees destined for Jamaica, among other islands, in addition to plants that were of interest to King George for his royal gardens [30]. Bligh would be elected to the Royal Society in 1801 but he never learned that command occasionally requires accommodation. As captain on another vessel he was court-martialed for abusive behavior but was acquitted in 1805. Again in 1808 as Governor of the Australian colony of New South Wales he was tasked with cleaning up a corrupt rum trade, but before long he was arrested by his own military for "crimes that render you unfit to exercise the supreme authority another moment in this colony". He was kept under guard for a year before being sent back to England [31].

The breadfruit did well in their new home, and were said to be hardy, prolific and easy to cultivate. Correspondence with Joseph Banks confirmed that 3–4 trees would be more than sufficient to feed a person for a year and that the fruit made a delicious pudding [32, 33]. There was only one drawback. Few of its intended consumers liked the taste, and breadfruit became quite unpopular among African slaves. In the meantime, from its arrival in 1791 breadfruit was not the savior it was intended to be. The natural disasters of the previous decade had abated, and trade with the United States resumed. There were now alternative comestibles. And perhaps because of its stigma as 'slave food', it was not until emancipation decades later that breadfruit would be accepted in its new island homes.

The Practice and Influence of Whaling

There is little information on hunting of whales and their relatives by Pacific islanders, but judging from harpoon heads that have been found in a number of island groups, it is likely that porpoises and dolphins were hunted. Harpoons of various types made from bone were lashed to a pole and were either flat or round but had two barbs [34]. In the Tuamotu Islands such animals were driven into shallow water with the sounds made by beating on canoe hulls [27]. Small sperm whale and pilot whale bones are common enough in some of the Society Islands to suggest that they were hunted, rather than simply being the remnants of beached specimens [35]. Large whales by contrast were considered scared by several Polynesian societies.

That was not the case among Europeans and Americans. Large whales were hunted beginning with the British in 1767, and the French in 1785, but the large whaling fleet from New England quickly supplanted them, particularly after the supply of Atlantic whales had been exhausted. At their peak in the mid-1800s there were no fewer than 700 whaling vessels plying the waters of the open Pacific. These were typically square-sailed ships (Fig. 4.9) about 30 m long with a crew of 30 or more men. The region stretching along the equator from the Line and Gilbert Islands, and south to the Marquesas and Society Islands were particularly favored for sperm whales that were there year round. The same was true for the waters near Fiji, Tonga and Samoa. Sperm whales elsewhere tended to be seasonal, but were hunted in the waters off New Guinea and the Solomon Islands nonetheless [36]. Mature males averaged more than 16 m in length but some were more than 20 m, with the head representing up to one-third of the body. Females were smaller, up to 12 m long. These large and powerful animals were speared by hand from 10 m-long harpoon boats rowed from the larger vessel (Fig. 4.9). This was dangerous work in which the small boat could be attacked or capsize when the injured whale made a deep dive to escape.

Fig. 4.9 The harpooning of sperm whale, oil painting by Thomas Beale in R. Hamilton, The Naturalists Library vol VII, Mammalia, Whales W.H. Lizars, Edinburgh 1843

Sperm whales that were successfully killed were towed and brought alongside the ship where a hole was cut between the eye and the dorsal fin so the animal could be impaled with massive iron hooks attached to block and tackle. Hard labor was required to force the whale carcass onto its side where the cutting process began. The lower jaw containing sperm whale teeth was separated from the body using large spade-like knives on 5 m poles; the head was then cut off with axes and hauled to the deck using additional blocks, tackles and hooks. The body, still aside the ship, was then cut spirally with similar spade knives to remove large strips of blubber. These were likewise taken on deck, cut into squares and placed into boiling water in large iron kettles that were set into a brick furnace on the deck for the rendering (extraction) process. The furnace, also called the 'try works' distinguished whaling vessels from all others, even those that were repurposed merchant vessels. Black smoke rose from the kettles, and as oil was released from the blubber it was manually scooped up with ladles and placed into casks to cool. This cutting, boiling and ladling process went on for six-hour shifts until the whale was entirely stripped. In the case of the sperm whale, the head was cut open to ladle out the spermaceti, oil of a superior quality that as a liquid wax was far more valuable than body oil and was used to lubricate precision instruments as well as for candles and cosmetic creams among other products.

All of this butchering was messy, bloody and macabre. The deck was slippery with oil and grease; soot from the boiling pots covered everything. After boiling, the blubber still had flesh and meat on them and these had to be extracted from the pots and placed in open casks to rot so that more oil could be extracted from the putrid mass. The decomposed flesh and blubber then had to be removed manually by the crew, a process that took hours and was undeniably nauseating. And even after cleaning the deck, the smell of a whale ship could travel considerable distances downwind from it [37].

The voyages were long and 1–3 years at sea was common. The typical whaling vessel was dark and dank below decks, and was typically infested with rats. The crew's rations included greasy pork, hard biscuits, heavily salted beef, pork, or horse, beans, rice, or potatoes. After months of hard labor and backbreaking work, and rations like these, the chance to obtain water, fresh fruits and vegetables at tropical ports of call was something to look forward to. The Marquesas Islands, Hawaii, Samoa and Tahiti were favorite stopover points for rest and recreation, which typically consisted of procuring alcoholic beverages and women. As put by Douglas Oliver [38], it would be intriguing to calculate how many kegs of rum were drained, how many wedding vows were broken and how many cases of venereal disease were transmitted in those ports. As many as thirty ships per year at stopped at each point and began a two-way trade that became particularly well established in Hawaii where islanders had the opportunity to buy cotton goods, hardware and firearms. Reciprocally, a commercial market was established in which businessmen from the mainland supplied food, tools, liquor and many more commodities for ships and their crews. Sail makers, blacksmiths, carpenters, laundries, bakeries, and many of the larger companies that later dominated the economy of the islands (see Chap. 5) got their start as merchants or suppliers to the whaling industry.

4.4 Trade with China

Tea for Sandalwood

The sandalwood tree is a rather unremarkable, bush-like plant that grows in dry forests of several high island groups in the tropical Pacific. There are several species, all in the genus *Santalum*, including *S. austrocaledonium* found in New Caledonia and Vanuatu, and *S. yasi* in Fiji and Tonga [39]. Smaller island groups including the Society Islands and the Marquesas in French Polynesia developed endemic species unique to those islands [40] and in the Hawaiian Archipelago each of the main islands had forests with sandalwood species unique to its mountains [41]. All sandalwood trees are root parasites, that is, they have special root extensions that take nutrients from the roots of certain other nearby plants with which sandalwood associates. In addition they grow slowly and require 30 years or more to reach maturity [40].

As trade began to open up with China toward the end of the eighteenth century, the English discovered tea and it rapidly became an indispensible part of the refined life, replacing the less genteel pint of ale as the national beverage. At the time, nearly all of the tea imported to England came from China and therefore trade items were in high demand. Sandalwood was one such commodity because its oily wood was highly prized in China where it was used as incense, medicine, and for carving. Thus began the search for sources. However, it was soon discovered that not all sandalwood was of equal quality. Those with red heartwood were generally considered the best, followed by yellow and then white. The most valuable came from India and certain Indonesian islands, but the newly discovered Pacific islands were generally more accessible to traders. In many cases 'sandalwooders' were the first Europeans to land on those shores and they appeared at frequent intervals. Not only did they map the coasts of many islands in the south Pacific, they also charted predominant winds, currents and reefs [42].

Sandalwood oil is found in the heartwood of the tree and it cannot be tapped like latex or maple sugar. Instead the tree is simply cut down and the heartwood is then extracted. No one was using the term 'sustainable harvest'. The sandalwood trade began in Fiji in 1804 and within 12 years essentially all of it had been cut. The trade then moved to Hawaii. The harvest there started in 1810 by trading silks, clothing and alcohol with King Kamehameha I who had his commoners harvest the wood and bring it to waiting ships. At the time there was a semblance of royal control over how much could be taken. However, after Kamehameha's death in 1819 the king's successors became introduced to the idea of trade on credit, that is, harvest with the promise of later payment. Thus began the wholesale pillaging of Hawaiian sandalwood, and by the mid-1830s all of the Hawaiian trees were gone. Likewise, after the discovery of sandalwood in the small Marquesas Archipelago (French Polynesia, see Fig. 5.4) in 1814, it took only three years to entirely remove all that were commercially valuable [42].

The last sandalwood stronghold was in Melanesia, particularly on New Caledonia and its surrounding islands, and in the New Hebrides group. When large quantities were discovered in the late 1820s, sandalwooders were eager to exploit it, but that would not be so easy as it may have been elsewhere. Foreigners were generally unwelcome in Melanesia; warriors were aggressive and strangers were typically regarded with suspicion if not outright belligerence. The widespread practice of cannibalism in the region was a further disincentive. Political units were smaller than in Polynesia and chiefs were less powerful. Should a chief befriend a trader, the enemies of his clan were always close by and ready to attack. An additional hurdle that frustrated and perplexed the early traders was the lack of interest by the islanders in anything Europeans had to trade or give as gifts. That included ironware, the value of which was immediately recognized and highly prized by Polynesians. Such items took longer to become of value in Melanesia. These behaviors were generally viewed as good reasons to go elsewhere, but sandalwood traders were persistent. They hired armed Tongan and Hawaiian laborers to cut sandalwood after reaching an agreement with local chiefs. There were accidents and incidents of aggression in the forests, but the traders stayed safely aboard ship during these forays. Slowly, the islands were relieved of their sandalwood trees, and despite exploitation that lasted a remarkable thirty years in the New Hebrides, the entire trade ended in the mid-1860s, less than fifty years after their discovery there [42].

Bêche-de-Mer

Trade with China also influenced the harvest of two marine animal products in the tropical Pacific. The first of these is commonly called a sea cucumber in English, bêche-de-mer in French: (literally 'sea spade'), bislama in parts of Vanuatu, and a

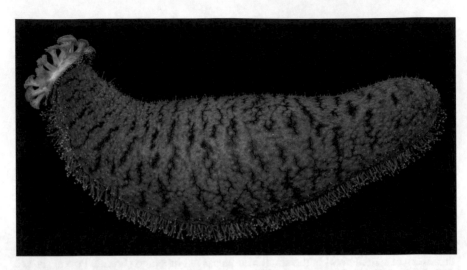

Fig. 4.10 A species of bêche-de-mer from Kosrae, Micronesia where it is called red fish. This particular species is found in relatively deep water adjacent to reefs where it uses its tentacles at the oral end (*left*) to scoop sediment and organic matter into its digestive tract. Tube feet below the body are used for locomotion as they are in their starfish relatives (Image courtesy of Wikipedia. org)

number of other names on different island groups. All of these refer to an echinoderm relative (e.g., sea urchins, starfish and allied groups). Sea cucumbers are bottom feeders that dig through and consume sediment (hence the term 'sea spade') for the organic matter it contains (Fig. 4.10). They are slow moving and easily captured. When prepared, they are cooked and the body wall, which has a muscular texture to begin with, is dried and then rehydrated, often in soups or in stews. On this author's palate bêche-de-mer had the consistency of cooked rubber with a taste to match, but apparently that opinion is not widely shared in east and Southeast Asia where many cultures consider them a delicacy. In some areas of the western Pacific and in Polynesia they are occasionally consumed as subsistence food, especially after cyclones, and on some island groups the intestines and gonads are consumed in addition to the body wall. There are at least 35 species that are harvested in the tropical Pacific, although some are more highly prized than others. The largest number of species is found in the vicinity of Papua New Guinea where many endemics occur. The number of species declines toward Polynesia and Micronesia from there [43].

The trade in bêche-de-mer began in Fiji in the 1820s almost immediately after the sandalwood rush had ended there. Ironware, trinkets and muskets were exchanged for them, but this required negotiation with Fijian chiefs to secure a labor force to comb reef flats and seagrass habitats at low tide. The animals were cooked over open pits that burned 24 h a day using timber that had to be cut and transported, sometimes over long distances. They were then gutted and placed in drying sheds. The work was laborious and dangerous, and was sometimes accompanied by attacks on the traders, an inherent risk in dealing with the 'Cannibal Isles' as Fiji was alternatively known. Populations of bêche-de-mer in shallow water were quickly depleted and the harvest boom entered a lull beginning in 1835. This allowed the population to revive, resulting in a boomlet lasting from 1842 to 1850, after which the trade did not recover [44, 45]. However, after a series of regulatory openings and closings in the last few decades, a small Fijian fishery for sea cucumbers persists.

The bêche-de-mer trade followed a similar boom and bust pattern in Micronesia, the Solomon Islands, French Polynesia and in Hawaii at various times during the 19th century. It continues as a fishery in some Melanesian regions following the increase in demand and price paid by Asian markets, although harvests open and close depending on fishing pressure. It is still exported to China from Tonga, Kiribati, New Caledonia and Fiji. At many localities, especially in Micronesia and Polynesia, the high value species have been and continue to be depleted [43].

Pearls and Pearl Oysters

Pearls are derived from the same lustrous and iridescent layers that line the inside of the pearl oyster shell. This inner shell is called 'mother-of-pearl' or MOP (Fig. 4.11) because it is the same material (sheets of flat calcium carbonate crystals) that the oyster places in concentric layers around an invader to produce a pearl. The irritant is typically a parasite, not a grain of sand as is often thought. However, while natural

Fig. 4.11 *Right*: open black-lipped pearl oyster from the Society Islands. The shell is anchored to the ocean bottom by threads from the byssus (on finger at *right*). The mantle tissue produces the smooth mother-of-pearl on the inside of the shell and also produces pearls. A black pearl (*lower left*) may be produced naturally but in this case it was produced as a cultured pearl using a shell implant (*upper left*) (Image (*right*) after Goldberg 2013 [48]. Images left courtesy of Wikipedia.Org

pearls are rare and have been supplanted by pearl culture techniques (see below), the shell is also used for a variety of purposes and is much easier to come by. The thick shells of the black-lipped oyster shell (*Pinctada margaretifera*) are fairly widespread in the western Pacific including sheltered waters around Papua New Guinea, the Solomon Islands, Fiji, and in the Northern Cook Islands, but they are especially common in the atoll-rich Tuamotu Archipelago of French Polynesia (see Fig. 5.4; see also Fig. 1.14). Here they are abundant in atoll lagoons where they attach themselves to rocks by means of strong protein threads called a byssus (Fig. 4.11) as is typical of pearl oysters generally.

The Tuamotuans were skilled free divers who harvested limited amounts of shell for their own use as jewelry, tools, fishhooks and lures. However, the potential value of this material was not lost on the early European explorers and by 1815 their ships were engaged in trading mother-of-pearl for export to China. By the 1840s spiraling prices led to a 'shell rush' in those islands, especially to produce carved objects, lacquer box inlays, fans, vanity cases and MOP buttons. The black-lipped oyster population quickly became depleted on shallow-water reefs and lagoons, and despite regulations intended to preserve stocks, by the 1880s fewer than half of the atolls that once produced oyster shell were doing so. Regulations then classified lagoons as closed, restricted or unrestricted with respect to pearl shell harvesting [46].

Certain oysters have a propensity for pearl production, although they are often small and irregular lumps that attract little attention from those who intend to profit from them. In order to command attention, pearls must have certain qualities including a diameter of at least 7–8 mm, often considered a standard size. Those that are 10–12 mm are often in the 'luxury' category and those that are larger than 12 mm are apt to be very expensive. However, size is not the only quality. The degree of roundness, the way in which the layers reflect light (the pearl's luster), and color are equally important. The typical pearl is more or less white, but some pearl oysters typically produce multicolor, gold, yellow, pink or black pearls. The color is partly a function of genetics, especially the color of the oyster tissue called the mantle that produces the shell and any pearls it may contain (Fig. 4.11), but color can also be influenced by water conditions and the oyster's food supply. Therefore even among black pearls there is considerable color variation that can range from green to combinations of black and gray.

The pearl oyster industry has been transformed from one that was dependent solely on the occasional wild catch to one that focuses on the culture of oysters, either taken from the wild or raised in hatcheries and then grown to maturity. Harvest may occur for the meat or the shell, but oysters that are intended for pearl culture begin with the surgical implant of a round bb-sized piece of shell. This shell nucleus originates from several species of freshwater mussels, especially the thick ones originating in the Mississippi River and its tributaries. Mussels are not oysters, and although they occasionally produce pearls, much of their commercial value is for implantation. The shells are cut into strips and then into cubes that are milled to their final rounded shape. Typically this work is done in Asia and then sent to

Polynesia and elsewhere. The implant is prepared with a coating of mantle tissue from a 'donor oyster' to minimize rejection by the recipient and to serve as the foundation for a 'cultured pearl'. These can be grown to commercial size in 1.5–3 years with at least roughly a millimeter of MOP deposited around the artificial nucleus (Fig. 4.11). The implantation procedure is a skill and sometime results in rejection by the host, or it may result in inflammation or lesions rather than a marketable pearl. Studies suggest that rejection is quite common and can account for nearly 50% of the implanted oysters [47]. Nonetheless this rate far outstrips the chances of finding a natural and marketable pearl.

Black-lipped oysters are now farmed commercially in parts of French Polynesia and the Northern Cook Islands, and experimentally or in trials in the Solomon Islands, parts of Micronesia and Kiribati. In 2004, French Polynesia was the top raw pearl producer with 26.9% of the total pearl market. For a long time Tahiti had a monopoly on the black pearl but competitors began to emerge at the end of the 1990s. As of 2006, there were a total of 516 pearl farms in 31 lagoons, mostly in the Tuamotu Archipelago. Ongoing research is intended to find the best genetic stocks and growth conditions for pearl production, which is estimated to bring in excess of $160 million dollars per year to French Polynesia [49].

4.5   Exotic and Invasive Species

The potential impact of exotic species, that is, introductions of plants and animals not native to a region, had yet to become of scientific interest. Indeed, what drives and controls some species to become harmfully invasive is still unpredictable (see **Text Box** below), although there are many cautionary tales of such introductions. Nonetheless, the Tahitians often took the animal gifts or trades as a new and interesting meal rather than for breeding, and that took care of some of the exotics. In other cases the long trans-Pacific voyage limited animal (and plant) survival, as did the heat of the tropics for which sheep, for example, were not particularly well adapted. Even the exotic species themselves were capable of eliminating one another.

> **Exotic Species** are not native to a region; they are outside of their normal range and are *introduced,* usually by humans, either deliberately or accidentally. Exotic, introduced, alien, non-indigenous and non-native species are all synonyms. Not all introduced species exhibit harmful environmental effects. Some simply fail to get beyond the establishment phase, while others become established but fail to spread or reach their reproductive potential.

Invasive species as used here are exotics that spread rapidly into a new region and have an adverse impact on native species. Stages of invasiveness may include non-native species that are localized and uncommon, widespread but uncommon, localized but dominant, or widespread and dominant. Invasive species in the latter category typically have wide environmental tolerances and high reproductive potential, features that are often exacerbated by lack of competition, predation or parasitism [50].

Large numbers of introduced dogs and pigs, for example, found the rabbits quite tasty and limited their survival on the island. Tahitian society itself often played a limiting role in the success of the introductions. If a chief decided that bitter citrus, pomegranates and quinces brought by Captain William Bligh (HMS Bounty) were of no value, they would fail to take care of them or simply have them destroyed. Conversely, Tahitian chiefs took to owning goats in the 1770s, not for their meat or milk, but as status symbols, and these animals quickly became established in Tahiti. It is unclear how these destructive herbivores were eliminated, but indeed they were during the 1820s.

Cattle, introduced at about the same time, were also unappreciated as sources of meat or milk by Tahitians, and instead were treated as curiosities. Cattle require grazing grassland and such areas existed in the Society group, but there was not a lot of it. If allowed into forested areas, cattle are especially destructive, often tearing plants from the ground, roots and all. Unlike in Hawaii where such introductions were successful because of royal attention to them, cattle were accorded little care in the Society Islands and as efforts to ensure their perpetuation or protection were minimal, they failed to flourish.

Cats and Rats

Not all introductions had ecologically happy endings. Domestic cats were useful as shipboard ratters, and when Wallis and Cook gave pregnant cats as gifts to Tahitian chiefs, the felines efficiently ate the small brown Pacific rat (*Rattus exulans*) that likely accompanied the first settlers. The cats also dined on the far larger black ship rat (*Rattus rattus*) and Norway rats (*Rattus norvegicus*) brought by Europeans. This was a good thing because all three rats were native bird predators and ate eggs and chicks from their nests as well. However, the cats were also fond of birds and quickly ate several flightless species to island extirpation or to complete extinction. For the many islands with endemic avian species, extirpation and extinction were the same phenomenon.

Cats are not present on most islands of the Pacific, but rats are, and sometimes cats and rats coexist. On Wake Atoll, for example, cats were introduced in the 1960s possibly to control Pacific rats, but as on Tahiti quickly became a problem for bird populations. Eradication of feral cats began in 2003 and when the process was completed the rat population exploded. This required rat removal using anticoagulant bait and traps. Unfortunately, the rat population has proven resilient for a variety of reasons, and efforts to remove them still continue [51, 52].

Pacific rats are probably now found on more islands in the Pacific than any other rat species. For example, they are found on 30 of the 35 atolls, in the Marshall Islands, and of the roughly 125 islands in French Polynesia, rats are now widespread on all but five small and uninhabited atolls. Some but not all rats have a long history of being introduced. Black rats invaded Clipperton Island, a remote uninhabited atoll in the Eastern Pacific after two large fishing boats were shipwrecked there in 1999 or 2000 and they appear to have invaded Futuna (central South Pacific) in 2007 or 2008, probably from a cargo ship [53].

Black rats are especially successful due to their ability to exploit virtually all tropical island habitats, the inability of other small rodents able to compete with them, and the absence of predators. They typically occur in forest or shrub land, but can exist on islands with minimal vegetation and even in mangrove forests, although they do not do well on dry islands where vegetation is sparse. Both Pacific and black rats can cause severe damage to native forests or crops. Sandalwood is especially vulnerable to seed predation, and coconut crops can be destroyed or damaged by black rats gnawing holes to access liquid and the fleshy fruit, as they have on Tonga, Fiji and in the Tokelau Islands [53].

In general plant material including leaves, stems and fruits form the basis of all rat diets. Insects often provide the protein. However, diets become altered when seabirds and turtles arrive for nesting. Rats, especially the larger black rat, consume young and small birds as well as eggs in the nest, along with turtle and tortoise eggs and hatchlings. It has been estimated that up to 2000 species of birds including ground-living rails, as well as pigeons, doves and parrots among others, have been eliminated by rat predation, assisted by the dogs and pigs brought by Polynesians and hunting by humans themselves [54]. There are ongoing projects to eradicate the rat population on several islands.

Arboreal Snakes

The island of Guam is the largest and southernmost of the Mariana Archipelago and the largest and most populous island in Micronesia with an area of 544 km^2 and a population of about 165,000. Its western shoreline faces the Philippine Sea and the Eastern beaches face the Pacific Ocean. The northern two-thirds of the island are primarily uplifted limestones with several volcanic high spots. Cliffs 90–180 m drop precipitously into the sea from the island's northern end. The southern features are essentially volcanic with an elongated mountain ridge dividing the inland valleys and coastline. The Spanish claimed the island in 1521 after

Magellan's landing, despite being settled by the Chamorro people for thousands of years. Guam was also a significant World War II battleground prior to the invasion of Saipan, and is now a focal point of travel to Asia and the islands of the western Pacific.

The remaining interior forests are a 'typhoon climax' community, limited to a ten meter-tall tree canopy. While birds have been the victims of rats, dogs and native peoples, another problem began sometime after WWII when a ship, likely from the nearby Admiralty Islands, inadvertently carried native brown tree snakes (*Boiga irregularis*) to formerly snake-free Guam sometime between 1946 and 1950 (Fig. 4.12). Within ten years from the initial sightings, the population exploded, possibly through intentional introductions designed to suppress a substantial rat population. The rats became less common, but after 1960 the snake population shot up to as many as 100 per hectare in some areas. This 1–2 m long serpent is arboreal, rear-fanged, and mildly venomous species to adult humans. It hides in trees among other places during the day, but becomes an active predator at night. It is believed to be responsible for hundreds of power outages, the loss of domestic and pet animals, the envenomation of human infants, the probable extirpation of native bat and lizard populations, and the loss of most native birds. The limestone forests of Guam have historically supported 25 bird species two of which are (were) endemic. One of them, the Guam flycatcher, was last seen in 1985 and is now believed to be extinct. The other, the flightless Guam rail, is extinct in the wild, but has been successfully bred in captivity off the island. Twelve species were likely extirpated as breeding residents on the main island, while eight others experienced declines of 90% throughout the island or at least in the north [55, 56]. An active trapping program is underway, but having nearly depleted the bird populations, larger snakes have been found scavenging garbage and even sneaking in to steal a hamburger off the barbeque [57].

Fig. 4.12 The brown tree snake, native to coastal Australia, New Guinea and its surrounding archipelagoes. It is not native to Guam but was introduced there shortly after World War II with devastating consequences to native species. Image courtesy of the U.S. Geological Survey and Wikimedia Commons

Exotic Ants, Plants and Plagues

There is a small group of ants called tramps that are particularly adept a stowing away on various modes of transportation and have now become worldwide pests. Indeed, at least four tramp ant species have been included in the International Union for the Conservation of Nature's 100 worst invasive species [58]. Possibly the 'worst of the worst' for the tropical Pacific is the yellow ant (*Anoplolepis gracilipes*). Crazy ants, as they are called, are noted for their rapid and erratic movements when disturbed or searching for food. Because the females do not fly off with a mate to form new colonies, fertile females live together and form massive 'supercolonies' of millions of individuals, not in nests or mounds, but rather in tree cavities and soil under stones or houses. As the sun sets they emerge and consume anything in their path. They have jaw structures (mandibles) that can bite, but their real contribution to pest hall of fame is their ability to squirt formic acid from the end of their abdomen. They specialize in squirting the noxious liquid into the eyes of their prey, which includes chicks, especially from ground nesting birds whose eyes become inflamed and swollen, impairing vision and even causing blindness. Domestic animals such as newborn chicken chicks, pigs and rabbits are also at risk. Supercolonies also have developed a propensity for digging around the roots of young coffee, coconut palms and sugar cane, exposing the roots, decreasing crop yields and increasing susceptibility to disease. They are almost everywhere on Pacific islands. In Melanesia they were first noted in New Guinea, New Caledonia and Fiji in the 1880s or earlier. By the early twentieth century yellow crazy ants were discovered in Vanuatu, French Polynesia and the Caroline and the Marshall Island groups of Micronesia [59]. Eradication programs by the US Fish and wildlife Service have begun on American islands that are prime seabird habitat on Palmyra Atoll and Johnston atolls [59]. Substantial population reductions are achievable but they are difficult and require dedicated work by the volunteer Crazy Ant Strike Team.

Plants can also become invasive. After the importation of guava to Tahiti by Captain Bligh, it became an island favorite and was widely planted for its fruit. However, within a few years it became a pest that grew everywhere and outcompeted both native plant species and even other exotics. It is still considered as one of the most invasive plants in Polynesia today [2]. Other imports from Europe included syphilis, gonorrhea, tuberculosis, influenza, typhoid, scarlet fever and smallpox. The European rats brought fleas and bubonic plague with them, which had devastating effects on the unprepared Tahitian immune system. Earlier population estimates were figured at 35,000–120,000, but after sickness and disease introductions, the human population of Tahiti plummeted to about 6000 by 1802 [60]. It may have been easy to conclude that the gods were displeased.

References

1. Walker LR, Bellingham P (2011) Island environments in a changing world. Cambridge University Press, Cambridge
2. Newell J (2010) Trading nature, Tahitians, Europeans and ecological exchange. University of Hawaii Press, Honolulu
3. Oliver D (2002) Polynesia in early prehistoric times. Bess Press, Honolulu
4. Thompson VA, Lebrasseur O et al (2014) Using ancient DNA to study the origins and dispersal of ancestral Polynesian chickens across the Pacific. Proc Natl Acad Sci USA 111:4826–4831
5. Oskarsson MCR, Klütsch CFC et al (2012) Mitochondrial DNA data indicate an introduction through Mainland Southeast Asia for Australian dingoes and Polynesian domestic dogs. Proc R Soc B 279:967–974
6. Cairns KM, Wilton AN (2016) New insights on the history of canids in Oceania based on mitochondrial and nuclear data. Genetica 144:553–565
7. Matisoo-Smith E (2007) Animal translocations, genetic variation and the human settlement of the Pacific. In: Friedlaender JS (ed) Genes, language and culture history in the southwest Pacific. Oxford, Oxford University Press, pp 157–170
8. Larson G, Albarella U et al (2007) Phylogeny and ancient DNA of *Sus* provides insights into neolithic expansion in Island Southeast Asia and Oceania. Proc Natl Acad Sci USA 104:4834–4839
9. Dobney K, Cucchi T, Larson G (2008) The pigs of island Southeast Asia and the Pacific: new evidence of taxonomic status and human-mediated dispersal. Asian Perspect 7:59–74
10. Miles WFS (1997) Pigs, politics and social change in Vanuatu. Soc Anim 5:155–167
11. Einzig P (1966) Primitive money in its ethnological, historical and economic aspects, 2nd edn. Pergamon Press, Oxford
12. National Tropical Botanical Garden. http://ntbg.org/breadfruit/breadfruit
13. Zerega NJ, Ragone D, Motley T (2005) Sytematics and species limits of breadfruit (Artocarpus, Moraceae). Syst Bot 30:603–615
14. Onwueme I (1999) Taro cultivation in Asia and in the Pacific. FAO. http://www.fao.org/docrep/005/ac450e/ac450e00.htm
15. FAO (2010) Pacific Food Security Toolkit 4 FAO.Org Pacific Root Crops. http://www.fao.org/docrep/013/am014e/am014e04.pdf
16. Weisler MI (1999) The antiquity of aroid pit agriculture and significance of buried A horizons on Pacific atolls. Geoarchaeology 14:621–654
17. Woodroffe CD (2008) Reef island topography and the vulnerability of atolls to sea-level rise. Glob Planet Chang 62:77–96
18. Kirch PV (1979) Subsistence and ecology. In: Jennings JD (ed) The prehistory of Polynesia. Harvard University Press, Cambridge, pp 286–307
19. Roullier C, Benoit L et al (2013) Historical collections reveal patterns of diffusion of sweet potato in Oceania obscured by modern plant movements and recombination. Proc Natl Acad Sci USA 110:2205–2210
20. Gunn BF, Baudouin L, Olsen KM (2011) Independent origins of cultivated coconut (*Cocos nucifera* L.) in the old world tropics. PLoS ONE 6(6):e21143. https://doi.org/10.1371/journal.pone.0021143
21. Stoddart DR (1968) Catastrophic human interference with coral atoll ecosystems. Geography 53:25–40
22. Harries HC, Clement CR (2014) Long-distance dispersal of the coconut palm by migration within the coral atoll ecosystem. Ann Bot 113:565–570
23. Lal BV, Fortune K (2000) The Pacific Islands, an Encyclopedia. University of Hawaii Press, Honolulu
24. http://www.theplantlist.org/browse/A/Pandanaceae/Pandanus/
25. Thomason LAJ, Englberger L et al (2006) *Pandanus tectorius* (pandanus). In: Elevitch CR (ed) Species profiles for Pacific Island agroforestry. Permanent Agriculture Resources, Holualoa. www.traditionaltree.org

26. Nelson SC, Ploetz RC, Kepler AK (2006) *Musa* species (bananas and plantains). In: Elevitch CR (ed) Species profiles for Pacific Island agroforestry. Permanent Agriculture Resources, Holualoa. www.traditionaltree.org
27. Torrente F (2015) Ancestral fishing techniques and rites on 'Anaa Atoll, Tuamotu Islands, French Polynesia. SPC Trad Mar Res Manage Knowl Inf Bull 35:18–25
28. Ellis W (1844) The history of the London Missionary Society, vol 1. John Snow, Paternoster Row, London
29. McCusker JJ, Menard RR (1991) The Economy of British America, 1607–1789. University of North Carolina Press, Chapel Hill
30. Powell D (1973) The voyage of the plant nursery, H.M.S. Providence, 1791–1793. Institute of Jamaica, Kingston
31. Dando-Collins S (2007) Captain Bligh's other mutiny: the true story of the military coup that turned Australia into a two-year rebel republic. Random House, Sydney
32. Frost A1 (1993) Sir Joseph Banks and the transfer of plants to and from the South Pacific, 1786–1798. Colony Press, Melbourne
33. Mackay D (1974) Banks, Bligh and breadfruit. New Zeal J Hist 8:61–77
34. Wallin P (1996) A unique find on Easter Island. Rapa Nui J 10:99–100
35. Leach BL, Intoh M, IGW S (1984) Fishing turtle exploitation and mammal hunting at Fa'ahia, Huanine, French Polynesia. Journal de la Société des Océanistes 40:183–197
36. Lever RJ (1964) Whales and whaling in the western Pacific. So Pac Bull 14:33–36
37. Verrill AH (1916) The real story of the whaler: whaling past and present. D. Appleton & Company, New York
38. Oliver DL (1989) The Pacific Islands. University of Hawaii Press, Honolulu
39. Thomson LAJ (2006) *Santalum austrocaledonium* and *S. yasi* (sandalwood). In: Elevitch CR (ed) Species profiles for Pacific Island agroforestry. Permanent Agriculture Resources, Honolulu. www.traditionaltree.org
40. http://ntbg.org/plants/plant_details.php?rid=1789&plantid=10201
41. St John H (1947) The history, present distribution and abundance of sandalwood on Oahu, Hawaiian Islands. Pac Sci 1:5–20
42. Shineberg D (1967) They came for sandalwood. A study of the sandalwood trade in the southwest Pacific, 1830–1865. Melbourne University Press, Carleton AU. Reprinted 2014, University of Queensland Press
43. Kinch J, Purcell S et al (2008) Population status, fisheries and trade of sea cucumbers in the Western Central Pacific. In: Toral-Granda V, Lovatelli A, Vasconcellos M (eds) Sea cucumbers. A global review of fisheries and trade, FAO fisheries and aquaculture technical paper no 516. FAO, Rome, pp 7–55
44. Campbell IC (1989) A history of the Pacific Islands. University of California Press, Berkeley
45. Fisher SR (2002) A history of the Pacific Islands. Palgrave, Hampshire
46. Rapaport M (1995) Oysterlust: islanders, entrepreneurs, and colonial policy over Tuamotu lagoons. J Pac Hist 30:39–52
47. Cochennec-Laureau N, Montagnani C et al (2010) A histological examination of grafting success in pearl oyster *Pinctada margaritifera* in French Polynesia. Aquat Living Resour 23:131–140
48. Goldberg W (2013) The biology of reefs and reef organisms. University of Chicago Press, Chicago
49. Tisdell C, Poirine B (2008) Economics of pearl farming. In: Southgate PC, Lucas JS (eds) The pearl oyster, a beginner's guide to programming images, animation and interaction. Elsevier BV, Amsterdam, pp 473–495. http://espace.library.uq.edu.au/view/UQ:155046/WP143.pdf
50. Colautti RI, MacIsaac HJ (2004) A neutral terminology to define 'invasive' species. Divers Distrib 10:135–141
51. Rauzon MJ, Everett WT et al (2008) Eradication of feral cats at wake Atoll. Atoll Res Bull 560:1–21

52. Griffiths R, Wegmann A et al (2014) The Wake Island rodent eradication: part success, part failure, but wholly instructive. Proceedings of 26th Vert Pest Conference, University of California, Davis, p 101–111
53. Harper GA, Bunbury N (2015) Invasive rats on tropical islands: their population biology and impacts on native species. Glob Ecol Conserv 3:607–627
54. Steadman DW (1995) Prehistoric extinctions Pacific Island birds: biodiversity meets zooar-chaeology. Science 267:1123–1131
55. Rodda GH, Fritts TH, Conry PJ (1992) Origin and population growth of the brown tree snake, *Boiga irregularis* on Guam. Pac Sci 46:46–57
56. Wiles GJ, Bart J (2003) Impacts of the brown tree snake: patterns of decline and species persistence in Guam's avifauna. Conserv Biol 17:1350–1360
57. Fritts TH, Leasman-Tanner D (2008) The brown treesnake on Guam. US Geol Surv. http://www.fort.usgs.gov/Resources/Education/BTS/
58. Lowe S, Browne M et al (2004) 1000 of the world's worst invasive species. www.issg.org/booklet.pdf
59. http://www.cabi.org/isc/datasheet/5575. Invasive Species Compendium *Anoplolepis gracilipes*
60. Finney BR (2007) Tahiti: polynesian peasants and proletarians. Transaction Publishers, Piscataway

Chapter 5
The Cultural and Political Impact of Missionaries and Foreign Hegemony in the Pacific Islands

Abstract This chapter examines elements of traditional societies of the tropical Pacific, emphasizing those of Polynesia and Tahiti in particular. In Tahiti and elsewhere in the Pacific the general sequence of events was that missionaries follow trade, and flags followed the missionaries. The (Protestant) London Missionary Society was the first to evangelize in these islands, and did so within 30 years of their European discovery. The interactions of the LMS with the native people and the cultural landscape they created are described, as are the political alliances they helped foster. However, Catholic France ultimately untied the relationship between the Tahitian monarchy and the British Protestantism. The British and the Wesleyan Church in Fiji became embroiled in a dispute between American business interests and a local king who was forced to protect his interests by asking Britain to annex his homeland. The role of plantation owners in the nefarious practice of blackbirding is also described. Missionaries additionally became active in the Hawaiian Islands in the early nineteenth century and gave rise to a business class that undermined the monarchy. This was accomplished through the actions of the political action of the Missionary Party, eventually causing the Kingdom of Hawaii to become a territory of the United States.

The discovery of the tropical islands of the Indian Ocean by the Portuguese has already been described, but they were more interested in the spice trade and hegemony around their colonies in India than they were in seeking converts to Christ. Although Catholic missionaries made conversion attempts, the Maldives remain an Islamic nation as they have been since the end of the twelfth century. Indeed, it is no longer possible to become a citizen of the Maldives without being Muslim.

The introduction of Western religion into the Pacific islands has an entirely different history. Again it was the Portuguese and the Spanish who encountered native western Pacific islanders in the late fourteenth and early fifteenth centuries (Chap. 3), and even though there were European designs on converting the heathen to Christianity during that time, little to that end was actually accomplished. Indeed, little was known of native people and their customs until the voyages of English

© Springer International Publishing AG 2018   111
W.M. Goldberg, *The Geography, Nature and History of the Tropical Pacific and its Islands*, World Regional Geography Book Series,
https://doi.org/10.1007/978-3-319-69532-7_5

captain Wallis (1767) to Tahiti, followed by James Cook. The islanders were generally friendly and as might be expected the crews of both vessels took liberties with the native women. French explorer Louis-Antoine de Bougainville visited the same area less than a year later and encountered the same hospitality except that a number of Bougainville's crew contracted syphilis. Thus began the reciprocal accusations of the French and the English for introducing venereal disease to paradise. Microbial afflictions notwithstanding, paradise is what both Wallis and Bougainville would later describe as the essence of Tahiti and the Society Islands generally- a Garden of Eden with bountiful fruit trees, friendly people, and a pleasant climate. And this was what brought Captain James Cook to the island repeatedly, as a focal point for his three cruises.

Some other islands also gave a relatively warm welcome to the Europeans. Tonga, for example, became known as the Friendly Isles. However, both Cook and Bougainville were aware that not all islands were equally receptive. The New Hebrides (now Vanuatu in Melanesia) had acquired a reputation for disliking outsiders as the early descriptions by de Quirós attest, and was similarly described by Cook as outlined below. Even in friendly Tahiti, Cook witnessed and described a ceremony that included human sacrifice in preparation for war with the neighboring island of Moorea [1] (Fig. 5.4). Inter-island warfare was commonplace here and elsewhere in the tropical Pacific.

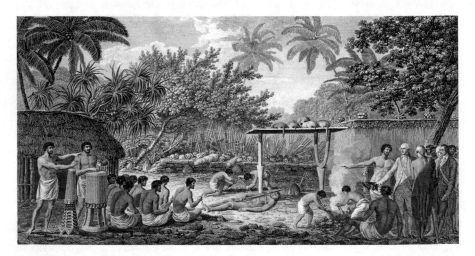

Fig. 5.1 Captain James Cook and his officers at right witness a human sacrifice being prepared for Oro in Tahiti. A grave is being prepared at the front of the stone altar, center accompanied by the beating of drums, *left*. Human skulls from previous sacrifices are shown on the stone altar in the background. A dog is cooked (at *right*) and offered with others (and pigs) on an elevated wooden altar. Engraving by William Woollett after drawing by the English artist John Webber who accompanied Cook (Image from an 1815 edition of Cook's voyages, courtesy of Wikipedia.org)

5.1   The Missionary Prelude: Polynesian Society and the Religious Pantheon

Polynesians worshipped innumerable and functionally malleable gods. They can be characterized generally as supernatural beings that were either active or inactive, but capable of becoming energized and potent. Active gods included those who had never been human, including Ta`aroa in Tahiti, the supreme god, creator of the universe, and of life and death. His shadow representatives on Earth were the whale and the blue shark, both of which were held in high esteem for that reason. An angry Ta'aroa might bring pestilence or draught that should dictate a human offering because the high gods liked human flesh. The sacrifices, with due respect for the high place accorded to the porcine, were sometimes referred to as 'long pig'. There were also humans who had become gods (e.g., ghosts of ancestors) and occasionally required pacification, or those who were descended from gods, including royalty. Other lesser gods, sometimes referred to as being 'departmental' played active roles in the affairs of living humans, controlling specific winds and ocean behaviors, fishing, harvest of the land, trickery, fertility and even thievery [2]. In Tahiti, Oro was god of both war and peace, perhaps representing a social continuum if not a contradiction. These gods who were active or who had suddenly become active, required their influence to be directed where it was desired and when the occasion required, the spirit sometimes had to be expunged from places where it was undesirable. Polynesians believed that all things in nature, including humans, contained a sacred and supernatural power, *mana,* that could be good or evil. Individuals, animals and inanimate objects contained varying amounts of mana, and because it was sacred, Polynesians had complicated rules to protect it.

Polynesian society was typically highly stratified. At the top were powerful chiefs, or sometimes a single king who oversaw many local chiefs. Because the aristocratic leader could trace his (or sometimes her) descent from the gods, royalty possessed more mana than other humans. The royal family had a separate residence from the rest of the population and dressed distinctively. The chief and his family also had tattoos that marked their royal status. The chiefly class was paid tribute in the form of food and material from the commoners and in addition had control over certain sacred forest areas and fishing grounds. These were marked as tapu (Chap. 2) and their boundaries were watched over by gods and spirit beings that crossed between the world of the living and the world of the dead. Due to the large number and wide variety of gods, Polynesian societies also developed a priestly class that served at the right hand of the aristocracy to manage communication between deities and royalty. Negotiating with gods was especially important during key periods of life, including conception, birth and death. Such events required careful selection of living gifts that could be sacrificed, allowing the priests to offer their spiritual essence [3]. Dolphins, turtles, pigs and dogs were the manifestation of spiritual beings and their mana was also held in high esteem. On portentous occasions they were offered along with humans (Fig. 5.1).

Below the kingly and priestly classes there were high-status warriors and other skilled specialists including master carvers, doctors, professional fishermen, musicians, and tattoo artistes. There was also a landowner class that oversaw the commoners, lower class workers who did not own property. Landowner holdings typically included family dwellings, a cookhouse, a garden, and places where domesticated chickens, pigs or dogs were kept.

Human Sacrifice

Cook described the altars on the Society Islands adorned with tikis of various sorts where sacrifices took place amidst numerous skulls of previous victims, along with offerings of pigs and fruits. Human sacrifices were typically selected from those who had offended the king or the priests, or who were captured in war. Families of those who previously had been sacrificed were considered tapu. They were enslaved and likely to be selected for future rituals. The victim was killed on the spot, usually by a blow to the head from a club or stone and then placed before a god carved from stone or wood. A priest typically removed one eye, placed it on a plantain leaf and presented it to the king or chieftain. The consumption of the eye is thought to be a symbolic substitute for eating the flesh [2, 4].

The body was not consumed but instead was wrapped in coconut fronds, and after decomposition, it was buried beneath the altar. That peculiarity was a Tahitian variation on a theme. On other islands including New Zealand and Fiji, the sacrifice was partly eaten. Indeed, the practice was so widespread and would become so well known among missionaries, that Fiji would later become known as the Cannibal Islands. Sacrifice and appeasement of the gods was occasioned not only by warfare, but also by pestilence, severe illness of the king, crop failures or other calamitous events. Indeed, failure to pay adequate homage to the gods was thought to be the cause of famine, epidemics, diseases and death.

Famine was especially prevalent on atolls and other low islands where impoverished limestone soils limited food to swamp taro, bananas, pandanus fruit and coconuts in addition to seafood. On Tongareva (now Penrhyn Atoll) in the Cook Islands 1300 km west of Tahiti for example, internecine warfare over food was devastatingly frequent. And because terrestrial food was even more limited on atolls than on other islands, wars were often fought over coconuts. The coconut trees of the vanquished would be cut down in an effort to limit their opponent's food. Thus while food shortages may have been the *casus belli*, the overall effect was to even further limit an already diminished food supply. That tactic must have made sense to those involved at the time. On other Cook Island atolls, such as Pukapuka, terrestrial foods were limited by the rain cycle to the first six months of the year. Famine was a recurrent problem in the following months. Food shortages also resulted from cyclones, especially in the cyclone belt a few degrees north and south of the equator. On Pukapuka a great tsunami in the 1600s divided history there into 'pre- and post-catastrophe' [5]. On the atoll of Ontong Java (north of New Guinea) by contrast, the two tribes inhabiting the islands existed relatively peacefully and lived on a diet primarily of taro and fish. Coconuts were available, but were used as a ceremonial ritual crop, and as a status food.

Sexuality

Nudity was the norm for young children and was not seen as being sexual, but rather an adaptation to a warm climate. Adult males wore a loincloth and adult females wore a leaf or bark skirt. The breasts were not covered. Once pubic hair began to grow, the organs of progeny were covered for protection; modesty or shame were concepts that were foreign to Polynesians [6]. The genitals possessed spiritual power, and were treated with respect and worship. Indeed, the religious aspects of fertility were expressed in song and sexually explicit ceremony as well as in the traditional woodcarvings of the powerful gods, whose genitals were prominently shown. During the traditional times in Polynesia, children were permitted open sex play and were able to observe adult sexual behavior due to one-room sleeping arrangements. On Tahiti, coital simulation became actual penetration as soon as young boys were physically able, and young girls may have engaged in copulation before age ten [7, 8].

Young boys were not circumcised, but superincised with a single cut made from front to back on the top of the foreskin by a priest. This was done at the age of 6 or 7 in Hawaii, but the Cook Islands it was performed at the onset of puberty. As the penis grew, the skin pulled back giving an appearance similar to true circumcision. This was done in the belief that the procedure increased sexual pleasure as well as fertility, and as an esthetic element, it made the penis look less like a hooded snake. The preparation for superincision was accompanied by blowing into the foreskin daily starting from birth. Then it was done less often, perhaps three times a week until the young male was 6 or 7. This task was often given to the "grandmother" or "aunt" who may have been a blood relative, a friend or a female neighbor. The process was said to loosen and balloon the foreskin and separate it from the glans, so that when the time came the skin was quickly and easily slit.

Attention was also paid to female genitalia for the same functional and esthetic reasons. While a female was still an infant, mother's breast milk was squirted into her vagina, and the labia were pressed together. The mons was rubbed with kukui (candlenut) oil and pressed with the palm of the hand to flatten it and make it less prominent. In some island groups, including several in Polynesia, the labia minor were stretched to make them longer and adult females similarly lengthened the clitoris orally. These practices related to the genitalia were not seen as being erotic, sexual, or abusive, but rather as an appropriate aspect of adult care of non-adults, and a necessary chore. Sex education came later. The training concerned not only what to expect and what to do but also how to increase or maximize pleasure. A young male was taught "timing" and how to please a female in order to help her attain orgasm. A young female was taught how to touch and caress a male and move her body to please them both. She was taught how to constrict and rhythmically contract her vaginal muscles, and some were able to make the labia 'wink' [8].

Although Polynesian societies condoned premarital sexual expression, wearing of perfumed oils, polygyny and polyandry, access to partners was strictly structured. In Polynesian societies that were highly stratified (for example in the Society Islands or in Hawaii) open sexuality among commoners was condoned, but among those of high status or royalty, virginity was valued and protected until a marriage with a partner of suitably high rank was arranged. On atolls where populations were often small, sexual relations among relatives were tapu and girls were brought to other islands to find a partner. The early European explorers took advantage of the casual and positive attitudes of Polynesia regarding sex, although the cultural structure and rules for sexual expression were not apparent to the Europeans. Instead, Polynesia became known for "erotic festivals, ceremonial orgies and sex expeditions" [8].

The Tattoo

Tattooing was practiced in Melanesia and in Micronesia, but it was most elaborately done in Polynesia where it was typically initiated after puberty. This art form served four purposes including a mark of physiological maturation, as a badge of courage and fortitude in battle, marks of personal and familial identity, and social status. The latter might include a specialized skill or material affluence. Men were much more adorned with tattoos than women and were often decorated from head to toe, although in some societies only men were tattooed. Markings also differed among island groups. Maori tattoos (New Zealand) called moko [9] included distinctive grooves left in the skin formed by chisels, a process that must have been excruciatingly painful compared with ordinary needle-puncture technique. These were expensive to obtain, and elaborate moko were usually limited to chiefs and highly ranked warriors (Fig. 5.2). Malu is a form of tattooing for females used in Samoa, which covers the legs from just below the knee to the upper thighs just below the buttocks. It is typically a finer and more delicate design compared with equivalent tattoos for males. Tattoos in some island groups as in the Marquesas Islands used designs that included animals, making them equally distinctive.

Infanticide

Infanticide was practiced in response to food shortages (induced for example by severe El Niño or La Niña conditions that caused excessive rainfall or dry periods).

Crop failures often resulted and this was especially critical on atolls where the types and amounts of food were limited. Infanticide appears to have been practiced for other reasons and varied among different island groups despite early missionary

Fig. 5.2 Maori chief
Tamaki Wata Nene with
intricate moko facial
tattoos (Painting by
Gottfreid Lindau (1890, oil
on canvas), Courtesy of
Wikipedia and the
Auckland Art Gallery)

reports that it was widespread and frequent. In stratified Tahiti and Hawaii, for example, infants born to parents from a mixture of upper and lower classes were strangled at birth. Infanticide was also used to increase the number of males in the patriarchal ranks [10, 11]. Abortion was also practiced, allegedly by using a blend of green pineapple and lemon juices that was available from older women specialists and was believed to cause miscarriage within two months [12].

5.2   The Arrival of The London Missionary Society in Tahiti

Against this backdrop, the missionaries arrived. A few Franciscan and Dominican friars successfully imported Catholicism to Spanish-held territory (The Mariana Islands and the Philippines) in the early 1600s. However, even though they became actively involved in evangelization, their ability to fan out from their posts was limited because the Jesuits lost favor in the courts of Catholic Europe, which grew suspicious of their power. By 1773 the Vatican ended Jesuit influence, and with it the Catholic Church's most prolific missionary organization [13].

That left the door open to the London Missionary Society, an amalgamation of Protestant groups composed of evangelical Anglicans, Calvinists, Baptists and

Congregationalists "not to send Presbyterianism, Independency, Episcopacy, or any other form of the Church Order, but the glorious Gospel of the blessed God to the heathen" [14]. They decided that their first objective would be Tahiti and Tonga in the South Seas. Polynesia generally and Tahiti in particular were chosen because success was considered more likely there due to the relative homogeneity of societies compared with Melanesia and Micronesia. The language was fairly simple and could be picked up in a few months, or at least that is what was thought. In addition, the political units were larger and chiefs or kings had more power there than elsewhere, a power that was exercised through religious observance. Therefore, conversion to Christianity might upend the entire political system in the favor of the Christian missions.

Tahitians must have been initially impressed with European sailing ships and their trade items such as glass beads, axes, spikes, large iron nails, mirrors and knives. Within a few years after the early voyages, trading exposed Tahitians to regular contact with non-missionary Europeans and Russians who brought with them cotton goods, rum and tobacco, as well as coveted hardware including muskets, powder and shot that could be used to advantage for carrying on traditional warfare. Then came the missionaries with the intention of staying, at least for a while. Their material wealth and skills opened a new world to the indigenous people, a world that became laced with the elements of desire and envy. Perhaps, they may have reasoned, in order to gain what the missionaries had, they would have to worship his God [15].

The missionaries left England in 1796. Among them were six carpenters, two bricklayers, two tailors, two shoemakers, a gardener, a surgeon and a harness maker. They were a collection representing each of 'the useful arts' intended be living examples to the natives. Seven months later they later they made landfall wearing tailcoats, high stockings, knee breeches and buckled shoes; their wives wore bonnets and heavy cotton skirts. Their instructions were to make friendly contact, build a mission house for sleeping and worship, offer examples of co-operative living, and learn the language of the island to be able to preach in the native tongue. On arrival they met King Pomare I, the royal decendant of the creator god Ta'aroa. Wearing a girdle of bark cloth, shark tooth and shell jewelry, and a crown of feathers, the king strode onto the beach accompanied by a slave crawling on his hands and knees. Pomare knew previous European explorers including Bougainville and Cook, and had provided Captain Bligh and the men of HMS Bounty a place to stay for the nearly six months they stayed in Tahiti gathering breadfruit trees for export to the Caribbean. Pomare accepted the missionaries as he had other foreigners, but he and the other natives watched them carefully. On one hand they were incredulous at the sight of them, while on the other hand they were carefully eyeing their possessions [14]. On settling in, the missionaries planted carefully tended gardens, which proved to be very attractive to pigs and goats. And because the missionaries were disinclined to share, the Tahitians often raided those gardens at night [3].

Attendance at church was not what the missionaries had hoped. Some Tahitians simply disappeared on Sunday, while others who attended fell asleep, talked among themselves, laughed and parroted the preachers' words back at them [3]. It was additionally unhelpful that some ruling chiefs made any church attendance a capital offence. In addition by that time, the trade in firearms and rum had already made Tahiti more dangerous than it had been before. With these disincentives, most of the first missionaries would soon be eager to leave.

Within a few years King Pomare I died and was succeed by his son who wanted to learn the 'talking marks' of books [14]. However, war broke out in 1808 and Pomare II was forced into exile on the nearby island of Moorea where the missionaries also fled. The young Pomare was clever and may have allowed them on his island to benefit his own leadership by stepping away from traditional gods and beliefs, thereby increasing his own influence. Perhaps if Pomare II could avoid reverence and deference to the unpredictable whims of Oro, he could assert his own power. It is not clear what his motivations were, but a growing number of islanders were coming to the mission in Moorea for religious instruction. The missionaries continued to play an important role in the affairs of island, and because affairs were inextricably linked to island war and peace, the role of the missionaries became political and not entirely limited to the work of converting the heathen.

Meanwhile Pomare II was eager to point out that he is beset with enemies and he is the only powerful friend the missionaries have. It would be a good idea for all concerned if his English friends might provide him with some military assistance. Indeed, it was with the assistance of missionaries that Pomare had already acquired swivel guns and a cannon from traders who came for salted pork (Chap. 4). And to supply his growing need for musket balls and similar ammunition, the missionaries allowed access to the iron in their storeroom. There were 800 converts on Moorea, all trained to use firearms. Under the appearance of religious services, Pomare and the missionaries kept them prepared for any hostilities [15]. In the meantime the missionaries themselves became involved in the pork trade, which resulted in a letter that appeared in the Sydney newspaper referring to them as 'gospel vendors and bacon curers' [3].

By 1819 Pomare II had been baptized and shortly thereafter his army converted to Christianity. They then invaded Tahiti and vanquished their enemies, a testament no doubt to the glory of God. The guns and ammo appeared to be equally helpful as did Pomare's belief in a new war god, Jehovah. But rather than slaughter the conquered and their families in the Polynesian tradition, Pomare protected them on the condition that they would agree to be baptized and converted to Christianity. With ardent missionary support, Pomare gained control of the entire island, regained the throne and enacted the Code of Pomare (1819), the first written law of Tahiti.

Interestingly, the Code was not merely a recitation of the Ten Commandments, but also included rules concerning the pork trade. No longer would Pomare allow missionaries or the agents of merchants on the island to act as middlemen. Now it would be illegal for anyone but Pomare himself to buy and sell pork [3].

Rather than attempt to combine the better elements of Tahitian society and some ancient customs with Christianity, the missionaries imposed a rigid interpretation of Protestantism. Under the Code only one religion was recognized, idols were burned, and tattoos as well as traditional arts and beliefs were banned. Cannibalism, of course, would have to go, as would polygamy, dancing and even singing anything

Fig. 5.3 Mother Hubbards carefully covering the skin of Tahitian girls, ca.1880s (Courtesy of Wikipedia.org and Museé de l'Homme and the French National Library)

but hymns. They were instead taught hard work, thrift, abstention, obedience, shame and modesty. Loincloths were substituted by trousers and long-sleeved shirts; leaves were replaced by Mother Hubbards (Fig. 5.3), sweaty, rain-soaked, loose-fitting dresses that covered women and girls from neck to toe, a somewhat imperfect attire for the tropics. The crusade against sin even prohibited wearing flowers in the hair. The code enforced strict observance of silence on Sunday and included penalties for such offenses as adultery, bigamy, theft, and rebellion.

The London Missionary Society had its first foothold in the South Seas and they began to consolidate their position by compiling a dictionary in the Tahitian language. This was not a simple task. Tahitian was not yet a written language and it had to be learned phonetically, including the exhaling sounds interrupted by glottal stops amidst its numerous successive vowels. Nonetheless, by 1838 a translation of the Bible was produced which greatly facilitated the conversion of the faithful. A Christian kingdom thus became established and spawned an army of native teachers that spread throughout the Society Islands [16]. It was a raging success, although at the considerable cost of suppressing native culture. A century later the English writer Robert Keable, who was also a vicar in the Anglican Church, would write that it was 'a thousand pities' the Tahitians had not converted the missionaries instead of the other way around [17].

The London Missionary Society enjoyed widespread success outside of the Society Islands, primarily as a result of their willingness to learn native languages and the use of local converts, which they called 'native agents', to evangelize in their own communities. However, their expansion into Samoa, Tonga and Fiji was only partly successful due to missions that became overextended financially and otherwise. In addition, there were conflicts with the Wesleyan Methodists (who were not part of the LMS) and wanted in on the evangelical action, especially in Samoa during the 1830s. Wesleyans did things a bit differently including the requirement of singing high praises of God while standing, and kneeling at prayer. Hymns, catechisms, class tickets and other features of religious expression were all different, but the Wesleyans claimed they had an agreement with the LMS to have free reign in Fiji in exchange for LMS dominance in Samoa. However, there were misunderstandings among the faithful, and the Wesleyans began sending teachers to Samoa. The LMS not only disputed the existence of any agreement, but also regarded the entire affair as "Wesleyan aggression and a breach of faith" [18]. LMS Calvinist missionary Reverend John Williams, who was noted for his successes in Tahiti, saw it this way: "The [Samoan] natives, though comprehending, but very imperfectly our objects, would at once discern a difference in the modes of worship, and their attention would of necessity be divided and distracted. Being also of an inquisitive disposition, they would demand a reason for every little deviation, which would then lead to explanations, first from one party, and then the other, and thus evils would arise, which otherwise might never have existed" [14]. Ultimately, the Wesleyans moved on from Samoa, but they were more successful elsewhere. In Tonga, for example, they account for 35% of the population and in Fiji they claim more than a quarter of the population today [19].

5.3   How the Kingdom of Tahiti Became French

King Pomare II was installed as the absolute ruler of Tahiti with the help of the London Missionary Society. But Pomare II did not rule for long as a consequence of alcoholism and Pomare's young son, Pomare III, died on his way to inherit his father's throne. That left his older sister who became Queen (Pomare IV) at the age of 14. The queen was educated as a Christian and was a staunch supporter of the English missionaries. However, within a few years of her reign the French (who were smarting from being outmaneuvered by the English) were interested in claiming territory in Polynesia as a means of extending their influence. In 1834 two Catholic priests arrived in the Gambier Islands 1800 km from Tahiti (Fig. 5.4), and when it was learned that they had exerted firm religious and secular control of those islands, the queen immediately outlawed Catholic missionaries. In the meantime, the priests from the Gambiers were quietly imported to Tahiti in defiance of the ban and when Queen Pomare was told, she had them escorted in the middle of the night and placed on a ship bound for Chile. The French constituency on Tahiti was outraged and asked for help from the mother country. It came in the form of an armed ship, which

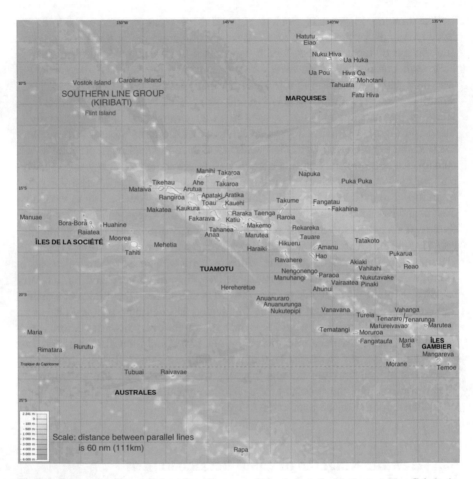

Fig. 5.4 Present-day French Polynesia with groups of islands described in the text. New Caledonia, 4000 km west of Tahiti, is not shown (Map courtesy of Wikipedia.org)

anchored in Tahiti's main harbor in September 1838. The Queen was presented with a series of unauthorized demands from the pompous Admiral Du Petit-Thouars, including a written apology to the French government, the reinstatement of Catholic missionaries, a cash payment, possession of one of the outlying islands and a twenty one-gun salute. The latter proved to be somewhat problematic as the Queen possessed only one cannon and not enough powder or shot for the job [20].

Claims and counterclaims, payments and apologies began to escalate until 1842 when the admiral made an encore appearance, this time with a squadron of ships bristling with cannons pointed at Pape'ete, the main settlement. This time, he took power. The queen was forced to yield as several hundred soldiers who were accompanied by French Catholic missionaries, disgorged from the anchored vessels. In France this action was nuanced as the declaration of a French Protectorate. In the meantime, the London Missionary Society representatives were, how shall we say, disinvited from further residence in Tahiti. Resistance to French rule began almost immediately as unhappy Tahitians fled to the interior and fought a guerilla action known as the French-Tahitian War. Although they evicted the French from Bora Bora and other islands with heavy losses on both sides, the Tahitians lacked guns and ammunition and the French prevailed. By 1846 France controlled Tahiti, Bora Bora Moorea, and later, the Tuamotu and Austral Islands, considered a dependency of Queen Pomare. However unhappy the English were with Tahitian behavior, they tended to respect island custom. However, the French were not nearly as accommodating. After declaring martial law in Tahiti they began to reshape the island's economy, regulated the pork trade, and established cotton, copra and vanilla plantations. And after fencing off large areas of grazing land contrary to Tahitian social norms, cattle were brought to French Polynesia where they can be found to this day [3].

Prior to French claims in Tahiti, residents of the Marquesas Islands (Fig. 5.4) about 1400 km northeast of Tahiti were being abused by whalers and traders in defiance of a local chief who asked for intervention. The French accommodated him with the same ships under the command of Admiral Du Petit-Thouars that would later visit Tahiti. Thus in 1842 the Marquesas became the first French Protectorate. However, things did not go so well farther afield on the Melanesian island of New Caledonia. In 1843, French missionaries there came to grief as dinner 'guests' while attempting to establish a foothold for Catholicism. This occurrence was far from unique and indeed was a repetition of attempts to bring Christianity to Melanesia in general as described below. Such setbacks notwithstanding, by the mid-1880s France began to affirm their grasp on all of these islands with the establishment of a governor and a general council. The islands that would later become the French Overseas Territory of New Caledonia and French Polynesia were now a colony [21].

5.4   The Misadventures of Missionaries in Melanesia

Emboldened by successes elsewhere, in 1839 the Reverend John Williams from the London Missionary Society decided to try his hand at converting the heathen in the New Hebrides Islands (now Vanuatu) in Melanesia. Captain James Cook visited the same location some six decades before and it did not go well for his landing party. Similarly, this first visit by missionaries to these islands was a mistake. Williams and his assistant were met on the beach by a local tribe, where they were clubbed to death (Fig. 5.5) and ritually cannibalized. For some time afterward the island of Erromango was known among Europeans as Martyr's Island. Descriptions written about the murders in the nineteenth century characterize them as revenge attacks for the kidnapping and murder of two boys, sons of a local chief, by Australian sandal-wood traders a few weeks prior. Inevitably a chain reaction of revenge and reprisals took place between natives and European traders, perhaps coincident with the arrival of the first missionaries, although corroborative evidence from other contemporary accounts to support this view is weak [22].

Fig. 5.5 The massacre of the lamented missionary. The Reverend John Williams of the London Missionary Society attempts to bring the Gospel to the native people of Erromango, New Hebrides, 1839 (From artist John Baxter, 1841, Courtesy of Wikimedia.org)

It may not always have been that way. The world as seen from the native perspective was limited to all that could be seen by the naked eye scattered in a vast sea. Everything ended at the horizon and plunged into the ocean. Nobody could venture beyond those boundaries and nothing was expected to enter. Ancestors departed to the world below the sea when they died. Thus, when the Europeans arrived they were thought to be returning ancestors, complete with pasty white skin as would be expected of ghosts [23]. It is easy to imagine the emotion and the excitement of the New Hebrideans upon seeing their relatives on the beach whom they welcomed with wild and exuberant dances that included raising their bows and arrows and clubs high above their heads. However, the returning "ancestors" including the Spanish expedition by de Quirós in 1606, followed later by the English and the French, were suspicious and on edge. And after being shot by arquebuses, muskets and pistols, as well as being abused, kidnapped and generally exploited by explorers and traders, the native people of the New Hebrides became reciprocally hostile toward white men [23]. There are, however, additional explanations. Fear of illness through sorcery is one. Early visits by Europeans often brought strange sicknesses with them, and murder of strangers may have been seen as a cure for pestilence. One such outbreak of sickness that occurred on Erromango prior to the visit by John Williams may have been the catalyst for his fate [22].

5.5   American Business and British Missionary Influences in Fiji

Fiji is part of Melanesia and consists of about 330 islands covering an area of 1.3 million square kilometers. Foreign interests in Fiji began in the early nineteenth century with the discovery of harvestable sandalwood trees (Chap. 4), and Wesleyan missionaries arrived shortly thereafter. In the 1840s American entrepreneurs started buying land from local chiefs to set up plantations and at that time businessman John Brown Williams was appointed consul by the U.S. State Department to over-see commerce and American interests. During a 4th of July celebration in 1849 overseen by Williams, the wadding from a saluting cannon struck and set fire to a Fijian hut. It spread quickly to Williams's property and destroyed his house, furniture, business and consular records. As the fire spread, the natives emptied the dwelling and later stole most of what they had salvaged from the flames. There were several other unfortunate incidents involving Williams and other Americans, which resulted in inflated claims of tens of thousands of dollars against a local Fijian king named Cakobau (Fig. 5.6), who of course was unable to pay. Williams threatened that the US Navy would collect on the debt and disenfranchise the monarch in the process [24]. Cakobau turned to the Wesleyan Missionary Society for help.

Fig. 5.6 Cakobau, a Fijian king (Photo by Francis H. Duffy circa 1870s, courtesy of Wikimedia.org)

Methodists had become very powerful in Fiji and by 1860 they had sixteen missionaries there, aided by 200 local preachers, more than 9000 lay members and 60,000 who at least occasionally attended church. Cakobau promptly converted to Christianity and offered Fiji to be ceded to Britain in exchange for payment of the debt while allowing Cakobau to retain his title. This offer was initially refused, which the missionaries regarded as evidence that Great Britain considered the responsibility of civilizing Fiji to be theirs. Indeed, they induced Cakobau to write a letter to Wesleyan headquarters in London to request more missionaries [25]. It did not escape the attention of the Methodists that there were attempts by the Catholic Church to 'infiltrate' some islands in Fiji and this caused what might be viewed as the anarchy of salvation. Indeed, the Methodist missionaries persisted in denouncing Roman Catholicism, and thought it their duty to do so. Native preachers in Fiji were instructed that 'popery' was contrary to the Bible. In several places the rivalry of religious sects led to acts of violence and to war in a few cases. On the little island of Rotumah, for instance, there was a pitched battle between the Wesleyans and the Roman Catholics as late as February 1871 in which six Roman Catholic natives including two chiefs were killed and many wounded [26].

Cotton, Sugar Cane, and Blackbirding

There is little doubt that the growing influence of missionaries in Fiji took the element of fear away from potential white immigrants, especially those who wanted to engage in the cotton business. During the American Civil War the Union blockaded southern ports and thus the export of cotton. Suddenly, the price of cotton began to skyrocket and European entrepreneurs thought that Fiji would be a perfect place to resupply the world with this product. Wesleyan missionaries in Fiji encouraged chiefs to plant it and within a few years cotton became a distinguishing feature of Christianized villages [27]. Europeans came to Fiji expecting to capitalize on the production of this fiber. However, after setting up shop they found that the locals were not only uninterested in the backbreaking work of cotton production, they were determined to be left out of it. And so, the plantation owners in Fiji decided to 'recruit' Melanesians from the Solomon Islands and New Hebrides to work the fields (Fig. 5.7). This was part of the 'blackbirding' trade described further in Chap. 6. Young males were especially valuable, and they were kidnapped, enslaved, and shipped to Fiji starting in 1865. This kind of free labor was not only useful for cotton, but also for European-owned sugar and copra plantations in Samoa and elsewhere in the Pacific. There was an increasing outcry from both the Wesleyans who were staunchly anti-slavery, as well as from humanitarians regarding labor practices and exploitation in Fiji. Traditional society had broken down so much that life and property were no longer secure, and Cakobau once again requested unconditional annexation with the assent of the other chiefs. The British took his offer and Fiji became a crown colony in 1874. It then became the headquarters of the British

Fig. 5.7 HMS Rosario from the British colony of Queensland in Australia intercepts the British slave ship Daphne in 1869 with 108 naked and chained men from the New Hebrides bound for Fijian plantations (Courtesy of Wikipedia.org)

Empire in the Pacific Islands for nearly 100 years. However, the cotton business there was short lived due to the early investors knowing little about tropical agriculture and having limited investment capital. Within a span of about 15 years cotton had become replaced by sugar cane production [27].

When the British Crown Colony was established, its first governor ended blackbirding and the sale of Fijian land to non-Fijians. He also encouraged sugar cane plantation owners to import laborers. Between 1879 and 1916 more than 60,000 indentured laborers were brought from India. Many stayed, leased land from the Fijians, and became small-scale farmers or raised cattle. Others became entrepreneurs, setting up shops in Fiji's urban areas. Many Fijians and Europeans, however, continued to view the Indians as second-class citizens, creating an animosity between the ethnic groups that exists today. The majority of indigenous Fijians today are Methodists, although the Hindu faith is represented by 28% of the population [28].

5.6 The Kingdom of Hawaii and How it Became American Territory

Captain James Cook visited the Hawaiian Islands in 1778 and named them the Sandwich Isles after his patron the Earl of Sandwich. Native Hawaiians welcomed Cook and crew as gods, but after they viewed a crewmember who had died, they became less in awe of the English. Relations deteriorated. It would be Cook's third and last voyage due to an altercation with the natives involving the theft of a small boat. Cook retaliated by attempting to kidnap a local king, and on retreat toward shore, Cook and four of his crew were killed at the hands of the Hawaiians [29].

Kapu, the code of sacred conduct, governed religious and civil society in Hawaii. These laws were similar to the tapu system of other Polynesian cultures (Chap. 2) and included no contact by commoners with Kings (chiefs). They were not allowed to touch any part of him including his cut hair or fingernails, or to trespass on royal areas, and were even forbidden to look directly at him. Men and women did not eat together, and women were not allowed to eat certain foods considered holy including pork, coconuts, and certain large fish. Violations of Kapu were often capital offences as dictated by tradition and perhaps by the mood or personality of the chief. However, the appearance of western traders and others who might give a monarch a political or military edge changed the nature of Hawaiian society, and by the early 19th century the powerful King Kamehameha I began to take an interest in foreign trade. Sandalwood exploitation was short lived (Chap. 4) but pineapple and coffee agriculture followed, as did foreign ships engaged in trading and whaling. Indeed Hawaii became the rest and recreation center of the whaling industry in the Pacific, and that magnetic status came with shipyards, boarding houses, saloons, and brothels. Whalers and merchant seamen came with certain proclivities for drunkenness, debauchery, vulgarity, and violence. They also came equipped with a microbial army of venereal and epidemic diseases including smallpox and measels that would leave their mark in Hawaii much as they had in other ports of call. The powerful King Kamehameha I had on one hand embraced kapu while reaping the benefits of trade on the other. Commoners, by contrast, were increasingly dismayed by tax burdens for which they received little. This dichotomy, and the social discord that developed from it, resulted in turmoil when Kamehameha died in 1819.

It was at this serendipitous time that the first missionaries, staunch New England Protestants, arrived in the Islands in 1820. They included seven freshly married Presbyterian, Congregationalist and Dutch Reformist couples from Boston. Marriage was a missionary requirement and some were so hastily arranged that four of the couples barely knew each other excepting their attendance at church, and they were married a month before the 3,000 km journey [30]. The new arrivals were appalled by Hawaiian customs including hula dancing, surfing, and kite flying, and promptly curtailed as many of those pagan activities as they could. Their objection to dancing was that it was sexually provocative as intended. They likewise frowned on kite flying as a wasteful violation of the Protestant work ethic, and the same stamp of disapproval was placed on surfing, not only because of the time it took from productive work, but because it was done in the buff while standing upright. The missionaries also did their best to discourage Catholicism brought to the islands by the French.

As the legitimate authority of kapu was being called into question, the latter-day royal family embraced the new faith. Missionaries served as royal advisors, protected women from abuse by traders and whalers, established schools, and were the first to provide a written language with consistent spelling for spoken Hawaiian. Chiefs found mana in learning to read and write in English and Hawaiian, and with the aid of the missions, a printing press in Honolulu produced a translation of the Bible and a legal code based on the Ten Commandments [31, 32].

By the 1840s the missionaries had accomplished much of their goals and while some went home or left for religious duty elsewhere, others stayed and established successful businesses and plantations. This led to their descendant's involvement in politics and thus the Missionary Party was born. The Hawaiian king at the time was encouraged to create a constitutional monarchy, but one that allowed privatization of land, including large tracts of American-owned plantations. Successor kings began to feel pressured by the political power wielded by the Missionary Party and attempted to reinstitute traditional Hawaiian culture that had gradually begun to fade. The Missionary Party viewed this renaissance as a regression to heathenism, and thus the Party plotted with American businesses and planters to disenfranchise the monarchy. In 1887, the Honolulu Rifles, a militia loyal to the Missionary Party, requested that Hawaiian king sign a new constitution. This offer was difficult to refuse, as it was made through the use of rifles with fixed bayonets. The Bayonet Constitution as it was called, led to the overthrow of the monarchy in 1893 and Hawaii was now declared a republic, the head of which was Sanford Dole, the son of a prominent missionary. Dole's government weathered several attempts to restore the monarchy, including an attempted armed rebellion. After several unrequited pleas for annexation, Hawaii was finally declared American territory and Mr. Dole would become its first governor in 1900 [33].

Due to the polyglot nature of its immigration patterns, religious expression in Hawaii today is not what the original missionaries might have expected. In the year 2000 about 45% of the population described themselves as religious, and of that number nearly half were Catholic, making it the largest denomination in the state. Protestants accounted for about 20% and 8% were members of the Church of Latter Day Saints. About 13% were members of eastern religions, primarily Buddhism [34].

5.7   Religious Competition on Pacific Islands

Polynesia is overwhelmingly Christian today and in many island groups the major players are Protestants who typically account for more than 50%, whereas Catholics are often 30% or more of the population. However, after the claim by France, the Marquesas Islands and New Caledonia became 90% Catholic. Likewise, the formerly Spanish colonies of the Mariana Islands (including Guam) and the Philippines are now 85% Catholic [35]. On some island groups the competition between

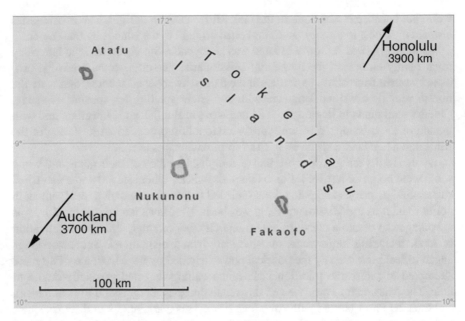

Fig. 5.8 The three atolls of Tokelau. Atafu and Fakafo are predominantly Congregationalist Christian but Nikunonu is almost entirely Roman Catholic. Map courtesy of Wikimedia.Org

Protestant and Catholics for converts became so intense that they split the population. On the three atolls of Tokelau midway between New Zealand and Hawaii (Fig. 5.8) for example, Atafu and Fakafo have been and are now predominantly Congregationalist Christian (>94% and >70% respectively) owing to the influence of the London Missionary Society, whereas Nikunonu, a small atoll 80 km or less from its neighbors is 97% Catholic [36].

There were other Christian groups that made conversion forays into Polynesia for their particular denomination. The success of the Wesleyans in Tonga and Fiji has already been mentioned. Representatives of the Mormon Church (Latter-Day Saints) held an interest in Polynesia not only to proselytize, but also perhaps to explore ancestral connections suggested in the Book of Mormon. LDS church members traveled to several islands in French Polynesia in the 1840s (in the Society and Austral island groups), but they were asked to leave by the French. They were more successful in Tonga where 17% of the population self-identify as LDS members, although the church claims a higher proportion [37]. More recently the Church of Latter Day Saints has also added to its membership in Kiribati and the Line Islands where they claim 10% of the population. Other religious groups with lesser representation may be found on various islands or archipelagos include Seventh-Day Adventists, Jehovah's Witnesses, the Salvation Army and the Assemblies of God (Pentecostal).

The practice of Christianity in its many forms often incorporated some traditional beliefs and mythologies including veneration of ancestors, references to spirits and the worship of icons. On Tahiti for example, voices and visions ascribed to Christian figures are instead attributed to traditional deities. In other adaptations, God is presented as Ta'aroa, the god of creation. By contrast. the variations of Christian faith in Melanesia are as varied as the number of their languages. In Tonga coconut juice and flesh are substituted for bread and wine in the Eucharist. The Anglican Church in the Solomon Islands allowed demon exorcisms, which continue to be conducted by the priesthood. Melanesian social elements such as warfare, sexuality, sorcery, magic and mysticism, generally referred to as Kastom in Pigin English, were integrated and incorporated by evangelizers into what they viewed as positive Christian traits. For example, the spirits of animals and plants could be interpreted as souls. Missionaries could also neatly fold the pagan notion that illness, wounds and death resulted from wrongful acts into the idea of original sin and the fall of man [38]. Rather than killing pigs at the cemetery during the funeral as a sacrifice to the spirit of the dead, pigs were killed and eaten while the elderly were still alive. The purpose of this shift was to emphasize communal relationships rather than the traditional attempt to make up for wrongs that may have been done in the past.

Such adaptations are reflected in denominational diversity. In Papua New Guinea, 96% of citizens identify themselves as members of a Christian church, but the denominations are Roman Catholic (30%), Evangelical Lutheran (20%), United Church (11.5%), Seventh-Day Adventist (10%), and Pentecostal (8.6%) [39]. Similarly, a high percentage of Vanuatu islanders self-identify as Christian, although the denominations are strongly influenced by where missionary groups were able to become established. Thus the northern islands are predominantly French Catholic, whereas the southerly islands tend to reflect the English influence of the Anglican and Presbyterian churches [16]. This division was not simply religious, but also became political as mistrust between European powers evolved into an unusual 'condominium' form of government that characterized Vanuatu for most of the 20th century (see Chap. 7, Table 7.2).

References

1. Synge MB (1897) Captain cook's voyages around the world. Thomas Nelson and Sons, London
2. Oliver D (2002) Polynesia in early historic times. Bess Press, Honolulu
3. Newell J (2010) Trading nature. Tahitians, Europeans and ecological exchange. University of Hawaii Press, Honolulu
4. Ellis W (1829) Polynesian researches fisher. Son and Jackson, London
5. Goldman I (1970) Ancient Polynesian Society. University of Chicago Press, Chicago
6. Sahlins M (1985) Islands of history. University of Chicago Press, Chicago
7. Gregerson E (1983) Sexual practices: the story of human sexuality. Franklin Watts, New York
8. Diamond M (2004) Sexual behavior in pre-contact Hawai'i: a sexual ethnography. Revista Española del Pacifico 16:37–58
9. Robley HG (1896) Moko; Maori Tattooing. Chapman & Hall, London
10. Tobin J (1997) Savages, the poor and the discourse of Hawaiian infanticide. J Polyn Soc 106:65–92
11. Goldman I (1955) Status rivalry and cultural evolution in Polynesia. Am Anthropol 57:680–697

12. Oliver DL (1981) Two Tahitian villages: a study in comparisons. Brigham Young University Press, Provo
13. de la Rosa AC (2015) Jesuits at the margins: missions and missionaries in the Marianas (1668–1769). Taylor and Francis, London
14. Hiney T (2000) On the missionary trail. Grove Press, New York
15. Senn N (1906) Tahiti the Island Paradise. WB Conkey Company, Chicago
16. Ernst M, Anisi A (2016) The historical development of Christianity in Oceania. In: Sanneh L, McClymond MJ (eds) The Wiley Blackwell companion to world Christianity, 1st edn. Wiley, West Sussex
17. Blond B (2005) Tahiti and French Polynesia. Transworld Publishers (NZ), Lonely Planet
18. Garrett J (1974) The conflict between the London Missionary Society and the Wesleyan Methodists in 19th century Samoa. J Pac Hist 9:65–80
19. http://www.state.gov/j/drl/rls/irf/2010_5/168355.htm. International Religious Freedom Report 2007: Tonga. United States Bureau of Democracy, Human Rights and Labor
20. Dodd ER (1983) The rape of Tahiti: a typical nineteenth-century colonial venture wherein several European powers with their Iron, Pox, Creed, Commerce, and Cannon Violate the Innocence of a Cluster of Lovely Polynesian Islands in the South Pacific Ocean. Dodd, Mead & Co, New York
21. Campbell IC (1989) A history of the Pacific Islands. University of California Press, Berkeley
22. Shineberg D (1967) They came for Sandalwood, beginnings of the trade. University of Queensland Press, Victoria
23. Codrington RH (1891) The Melanesians: studies in their anthropology and Folklore. Oxford University Press, London
24. Peabody Essex Museum. (1956). The life of John Brown Williams. The New Zealand Journal, 1842–1844 of Salem, Massachusetts. http://nzetc.victoria.ac.nz/tm/scholarly/tei-wilnewz-t1-g1-t1-body-d1.html
25. Darch JH (2009) Missionary imperialists? Missionaries, government and the growth of the British Empire in the Tropics, 1860–1885. Wipf and Stock Publishers, Eugene
26. Henderson GC (1931) Fiji and the Fijians 1835–1856. Chapter 13: Wesleyan Methodists and Roman Catholics. Angus &Robertson, Sydney. http://nzetc.victoria.ac.nz/tm/scholarly/tei-HenFiji-t1-body-d13.html
27. Stokes E (1968) The Fiji cotton boom in the eighteen sixties. New Zeal J Hist 2:165–177. http://www.nzjh.auckland.ac.nz/docs/1968/NZJH_02_2_05.pdf
28. 2007 Census, Fiji Bureau of Statistics. "Population by Religion and Province of Enumeration"
29. Collingridge V (2003) Captain Cook: the life, death and legacy of history's greatest explorer. Ebury Press, London
30. Grimshaw P (1989) American missionary wives in nineteenth-century Hawaii. University of Hawaii Press, Honolulu
31. Bingham H (1855) Sandwich Islands: civil, religious and political history of those islands. HD Goodwin, Canadaigua/New York
32. Fortune K (2000) Hiram Bingham. The Pacific Islands: An Encyclopedia, ed. by BV Lal and K. Fortune. University of Hawai'i Press, Honolulu
33. Schweizer NR (1994) His Hawaiian excellency. The overthrow of Hawaiian Monarchy and the Annexation of Hawai'i. Peter Lang Publishing, Bern
34. Roof WC, Silk M (eds) (2005) Religion and public life in the Pacific region, fluid identities. AltaMira Press, New York
35. http://www.worldpopulationstatistics.com/guam-population-2013/
36. Statistics New Zealand (2012) Profile of Tokelau Ata o Tokelau: 2011 Census of Population and Dwellings / Tuhiga Igoa a Tokelau 2011 mo te Faitau Aofaki o Tagata ma na Fale.
37. http://www.refworld.org/docid/519dd47f1d.html
38. Douglas B (2001) Encounters with the enemy? Academic reading of missionary narratives on Melanesians. Soc Comp Stud Soc Hist 43:37–64
39. http://www.state.gov/j/drl/rls/irf/2007/90150.htm

Chapter 6
Pacific Islands and the Politics of Fertilizer

Abstract Seabirds by the millions inhabit the islands of the tropical Pacific, especially in the equatorial regions of Peru where their phosphate-rich excrement (guano) has accumulated for thousands of years. The discovery of guano's use as a fertilizer in the 1840s led to commercial mining of it, especially along the coast of Peru where it had dramatic economic, social and political effects, including the use of slaves in the mining industry, declarations of war, and permanent border changes. The United States responded to its essential exclusion from this market by the approval of the Guano Islands Act of 1856, which enabled American business interests to claim uninhabited Pacific Islands with guano deposits. This led to claims and counterclaims between businesses as well as among competing countries. The island landscape changed on many island groups along with issues of ownership. The end of the guano trade was ushered in by the discovery of superphosphate deposits on certain isolated islands, and as the resource was being exploited, the strategic position of islands across the Pacific was becoming evident as the world prepared for war.

6.1 Seabirds and Guano

Seabirds tend to be abundant on isolated islands where they have access to food, an area to seek rest and refuge, and a place to nest. On isolated Pacific specks of land such as Midway Atoll (named for its position midway between San Francisco and Tokyo), various species of gannets, terns and albatrosses comprise a population of three million birds that occupy three small islands totaling roughly 600 hectares. That works out to about 5000 birds per hectare and a lot of bird poo. This mid-Pacific island is now a bird sanctuary operated by the National Wildlife Refuge, and it has succeeded despite having been dramatically altered by dredging to accommodate large vessels, and paving for airfields between 1938 and 1940. But how Midway and other islands similar to it became U.S. territory in the first place is an interesting story that requires a little background starting with Peru.

There are millions of seabirds, mostly pelicans, cormorants and boobies (Fig. 6.1), off the coast of Peru that feed on a small species of anchovy found in primarily along its coast called the anchoveta (*Engraulis ringens*). In good years,

W.M. Goldberg, *The Geography, Nature and History of the Tropical Pacific and its Islands*, World Regional Geography Book Series, https://doi.org/10.1007/978-3-319-69532-7_6

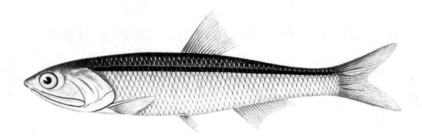

Fig. 6.1 Enormous schools of the Peruvian anchoveta, *Engraulis ringens*, each typically 10–25 cm long, support many predatory species including large populations of coastal seabirds (Drawing courtesy of Wikipedia.org)

Fig. 6.2 A crowd of birds on a Peruvian guano island dominated by guanay cormorants (*black plumage*), along with Peruvian boobies (*white plumage, lower right*) (Photo courtesy of Weimerskirch et al. 2010, PLoS ONE https://doi.org/10.1371/journal.pone.0009928)

favorable winds push surface waters away from the shore, forcing cooler and deeper waters to rise to the surface, a process called upwelling.

The upwelled waters bring nutrients to the surface, especially phosphate and nitrate, and these elements provide the foundation for phytoplankton and zooplankton growth. When upwelling is active off the Peruvian coast, as it normally is for 8–9 months of the year, billions of anchoveta are supported by enormous quantities of plankton, and seabirds, which in turn feed on the anchoveta. This cycle has been typical of the Peruvian coastline for thousands of years. And because seabirds prefer offshore islands near their food source (Fig. 6.2), the islands near the coast of Peru have accumulated these droppings. Millions of birds on the Chincha Islands south of Pisco left particularly rich deposits of guano nearly 50 m thick. Guano is a word derived from 'huano' in the indigenous Andean Quechua language, which refers to any manure that can be used as fertilizer. However, it became synonymous with bird excrement that is a bright and pasty white due to ammonium oxalate and its unusually high content of uric acid, both of which are sources of nitrogen. In addition, it contains a considerable amount of phosphate and potassium, among other minerals.

The value of bird poo as fertilizer was no secret to the peoples of Peru. They had been applying it to their crops for more than 1000 years and protected the birds under penalty of death for disturbing or killing them [1]. But while there were many such bird islands elsewhere, the three Chincha Islands, about 20 km from Pisco, produced guano that would turn out to be unique, and not only for the amounts of it. All guano contains three principal elements that can replenish nutrient-depleted soils. Phosphorus stimulates early root development, facilitates the formation of buds, flowers and the development of seeds. Potassium fosters a rapid maturation process, regulates metabolic activities and increases disease resistance. Nitrogen promotes the formation of new tissue and increases plant growth, and there was a lot of nitrogen in Chinchan guano due to its high uric acid levels (up to 16% by weight). The Chinchas are in the tropics, but they are in a particularly dry portion due to the rain shadow produced by the Andes Mountains. As a result the guano accumulated with little rainfall and low humidity that otherwise would have caused it to break down and release the important elements, especially nitrogen. This high-nitrogen characteristic gave Chinchan guano the stamp of quality by which all others would come to be judged [1].

European and American farmers were already experimenting with various animal and (untreated) human manures, ashes, bones, dried blood and other potential plant nutrients in the 1820s and 1830s. However it was not until the publication of a treatise by the German chemist Justus von Liebig in 1840 that the benefits of fertilizer use on crops got widespread attention. Liebig found that the soils supporting many crops were limited by nitrogen and phosphorus as illustrated by a barrel with short and long staves (Fig. 6.3). His 'Law of the Minimum' suggested that nutrient elements present in the least quantity would run out first as illustrated by nitrogen, the shortest stave. The water in the barrel, representing crop yields, cannot be increased without adding more of that limiting element. The next lowest stave, phosphorus, represents the subsequent element most likely to run out. Liebig used his knowledge of chemistry to help produce the first manufactured nitrogen-based fertilizer. However, as Liebig's work was becoming widely translated, the enriched element content of guano in the Chincha Islands was discovered, resulting in a boom for Peru's economy that would last more than 30 years. This was the Golden Age of Guano.

It may be worth mentioning that while Liebig's Law seemed to work on nutrient-depleted soils, results based on fertilizer effects alone do not take into account several other important factors that are now widely recognized. For example, alternative crops differ in the rates at which N or P may be used. Likewise, high concentrations of one element can influence the uptake rate of other elements. In addition, trace elements such as magnesium, manganese and iron can limit photosynthesis and can limit plant production. Indeed, any environmental factor such as water or light

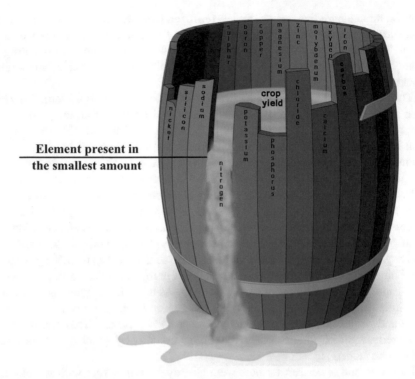

Element present in
the smallest amount

Fig. 6.3 Illustration of Liebig's Law of the Minimum from the nineteenth century German chemist Justus von Liebig. Barrel staves of different heights refer to quantities of plant nutrient elements. In this case, nitrogen is present in the least quantity and was thought to limit crop yield, illustrated by the amount of water the barrel can hold. His 1840 treatise was soon published in multiple languages and emphasized the importance of fertilizer in soil rejuvenation and the pivotal role it plays in increasing crop yields (Illustration modified from Wikipedia.org)

availability, temperature, or competition from other plants can limit plant growth. When two or more limiting factors interact they are said to act synergistically, and Liebig did not recognize what ecologists now refer to as 'factor interaction' [2, 3].

Guano From Peru: Not Much of It for You

Peru declared independence from Spain in 1821 but wars dragged on for another 5 years before the last royalists surrendered. Those conflicts were not the end of the independence movement, however, and were followed with later attempts by Spain to re-assert influence over Peru, including the occupation of the Chincha Islands and the blockade of Peruvian ports. There were fifty changes in government between 1821 and 1845, seven constitutions, political violence and civil war, all of which was followed by a crushing debt default that left the country bankrupt. However, in the 1840s the failed state of Peru was suddenly deemed creditworthy. Loans were made primarily from London banks to coordinate sales, shipping, warehousing and

THE CHINCHA (GUANO) ISLANDS: MIDDLE ISLAND, AS SEEN FROM NORTH ISLAND.

Fig. 6.4 Ships contracted by Great Britain to load guano on the Chincha Islands of Peru wait their turn in the harbor. It often took 2–3 months to load a single vessel (Image courtesy of Manuel González Olaechea y Franco and Wikipedia.org)

marketing expertise that Peru lacked. Why would London be willing to invest in such a highly unstable government? The short answer is guano, and that commodity allowed Peru to contract with a highly reputable British merchant house that would manage the country's guano exports and simultaneously service its foreign debt with the proceeds. For 25 years Peru was able to manage its finances, often by paying older debt with new loans. In fact, with its new guano economy Peru became the largest solvent debtor in Latin America [4].

Although Peru set the price for Guano, the British controlled a large share of its sale and distribution (Fig. 6.4). This served as a great annoyance to the United States because having to purchase Peruvian guano from the British was insulting. Even worse, as a third party purchaser, it was becoming increasingly expensive as the worldwide demand for guano rapidly exceeded the supply. This led to an explosion of American entrepreneurs to search for alternative sources including African, Arabian and Caribbean bird islands, but the guano from those areas was chemically inferior to that of the Chinchas. American farmers asked their representatives in Washington to do something-anything- about bringing down the price of guano, but Peru said no to any attempts at 'guano diplomacy' under the Fillmore administration [1].

In 1851 three bird-rich islands called Los Lobos were discovered more than 80 km from the Peruvian coast and other than the avian residents they were uninhabited. That would have perfect except that Peru claimed those islands, and was prepared to defend them despite being many times farther away from the widely recognized three-mile territorial limit. Los Lobos were dry islands, but not as dry as the Chinchas and the relatively high humidity there diminished the comparative quality of the guano. Nonetheless American entrepreneurs attempted to lobby for 'private bills' in Congress that would not only reimburse the considerable sums that had been spent in outfitting ships, but would also ensure protection of the claimants by the U.S. Navy when they went to Los Lobos. However after an elongated process of consideration and confusion within his administration, President Millard Fillmore decided that he would have none of it. There would be no U.S. backed confrontation over guano and Los Lobos were going to remain Peruvian. Perhaps there were other ways of finding a source of decent fertilizer.

6.2   The Guano Islands Act

By the mid-1850s there had been nine international incidents regarding attempts by Americans to take bird poo from islands of Peru, the Caribbean, and the central Pacific. None of these places belonged to the U.S. but that was a mere technicality and it was important to find a way around Peru's monopoly of the guano trade. As Senator William Seward put it before he became better known for purchasing Alaska, guano is the best fertilizer that is now known, and is a great article of commerce for American farmers, especially those tending worn out lands. Seward offered a bill to favor, encourage and protect American discovery and the discoverers who were placing themselves and their investments at risk for the benefit of the American people. It probably did not hurt the cause that several entrepreneurs had formed the American Guano Company that was reportedly capitalized to the tune of $10 million (Fig. 6.5).

Congress then passed the Guano Act and president Franklin Pierce signed Seward's bill into law in August of 1856. It enabled U.S. citizens to take temporary possession of islands with guano deposits as long as the islands were unoccupied and were not under the jurisdiction of another government. Interestingly, 'another government' included those from Europe and Great Britain, but did not apply to indigenous residents, although most claims involved uninhabited locations. The Guano Islands Act was one of many expansionist expressions of American might, and although the U.S. Supreme Court judged the Act constitutional (Jones vs. the U.S., 1890- see below), it was a law that was extraordinary and unique among nations. It allowed and indeed encouraged any private citizen with the means to claim territory for his country, as long as it had exploitable fertilizer.

In the years after passage of the Guano Act, 94 rocks, islands, cays and atolls were claimed in one fashion or another. The officially sanctioned method required more than just planting the flag, although that was often the extent of the claim. To be legally empowered under the Act the petitioner was required to demonstrate that the island had guano deposits. The island then had to be occupied by the claimants, and the application for title had to be bonded with the State Department. These requirements were not always met. In several cases, claims were made to uncharted islands that were so terribly misplaced they could not be found again. Companies

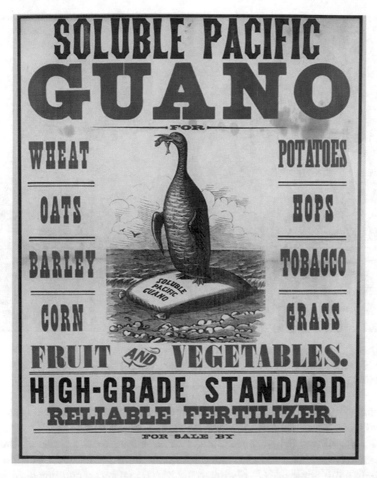

Fig. 6.5 Advertisement from the American Guano Company, ca. 1860 with cormorant and captured anchoveta in its beak. The ad was meant to suggest that this guano is Peruvian (or if questioned, that it was 'just as good'), but it was more likely mined at that time from the central Pacific islands and was comparatively deficient in nitrogen (Courtesy of the Mystic Seaport Museum)

also were accused of simply claiming every island in a group according to questionable whaling ship charts without demonstrating the presence of guano, or even landing on them to make note of the claim. For example, 47 islands scattered across the central Pacific were claimed on the same day, the 12th of February of 1859, by the U.S. Guano Company but only 23 of those places actually existed [1]. In other cases mining operations procured what appeared to be guano, but what was loaded aboard was later determined to be gypsum, a calcium sulfate used to make plaster of Paris. The 1200 tons of it brought from the central Pacific to New York by the American Guano Company was barely worth the cost of the voyage. After such incidents, businesses began to appreciate that some knowledge of the sciences was a virtue worth paying for.

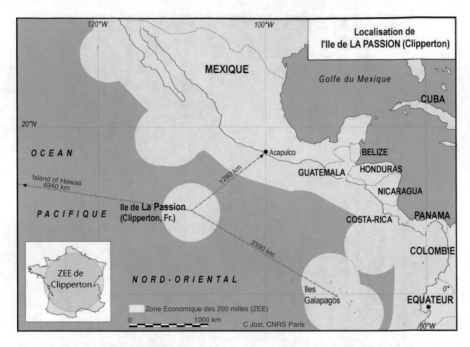

Fig. 6.6 Position of isolated Clipperton Island claimed by Mexico and by France. The nearest portion of French Polynesia is the Marquesas Islands located 4000 km SW of Clipperton (not shown) (Map courtesy of Wikipedia.org)

Multiple nations often became mired in claims made under the Guano Act. Isolated Clipperton Atoll in the eastern Pacific, for example, (Fig. 6.6) may have been named after a British pirate, but it was claimed by the Oceanic Phosphate Company of San Francisco and mined for guano during the 1890s. Mexico apparently took a dim view of U.S. claims to the island and sent a contingent of marines who unceremoniously hauled down the American flag. To make matters worse, the island was not bonded with the State Department as required, nor was it occupied or officially listed as U.S. territory under the Guano Act [5]. As a means of curtailing the discussion with Oceanic Phosphate, Mexico garrisoned troops on Clipperton, and then collaborated with the British Pacific Island Company in 1906 to mine the remaining guano. But the French who discovered the island in 1709 and named it the Île de la Passion, also claimed it as part of French Polynesia. The claims between Mexico and France were eventually submitted to international arbitration under the aegis of the Italian government, which pronounced *presto con brio* in 1909 that Clipperton was indeed French. Confusing as the intersecting claims were, French territory is the way it remains, but on most charts, sometimes even those in French, the island is still named after the swashbuckling English privateer, John Clipperton.

Of the claims made and the bonds posted by U.S. Guano and other American companies in the Pacific under the Guano Act, 35 were islands that actually existed (Table 6.1). These included several atolls in the Phoenix island group (Fig. 6.7) including Enderbury, McKean, Nikumaroro, Rawaki, Orono, Canton, McKean and

Table 6.1 Pacific Island acquisition and claims under the Guano Islands Act of 1856
Of 94 islands, cays, rocks and atolls under claimed the Guano Act, 66 were temporarily recognized by the State Department, although few of them were actually mined. The following list of 35 focuses on Pacific island claims (1857–1892) made under the Act. Claims made for Kiribati, Tokelau and Cook Islands were all counterclaimed by Great Britain. The eventual disposition is indicated by current affiliation. *Asterisks* indicate islands where guano was mined (Data from Skaggs (1994) [1])

Island Name	Original Name	Current Affiliation
1. French Frigate	"	Annexed by U.S. with Hawaii
2. Johnston*	Cornwallis	Unincorporated U.S. territory
3. Kingman	Danger	"
4. Palmyra	Samarang	"
5. Baker*	New Nantucket	"
6. Howland*		"
7. Jarvis*	Bunker	"
8. Caroline	"	Line Islands, Republic of Kiribati
9. Kiritimati	Christmas	"
10. Malden*	Independence	"
11. Starbuck	Barren	"
12. Tabuaeran	Fanning	"
13. Teraina	Washington	"
14. Vostok	Staver	
15. Flint*	"	
16. Birnie	"	Phoenix Islands, Republic of Kiribati
17. Kanton*	Canton	
18. Enderbury*	"	"
19. Manra*	Sydney	"
20. McKean*	"	"
21. Nikumaroro*	Gardner	"
22. Orona*	Hull	"
23. Rawaki*	Phoenix	"
24. Huon	"	Annexed by France with New Caledonia
25. Nuku Hiva	Madison	Marquesas Islands, France
26. Atafu	Duke of York	Tokelau Islands, New Zealand
27. Olosega	Swains	Unincorporated Territory, American Samoa
28. Nukufetau	Duke of Clarence	Republic of Tuvalu
29. Pukapuka	Dangerous	Cook Islands (affiliate of NZ)
30. Nassau	_____	"
31. Manahiki	Humphrey	"
32. Penrhyn	Tongareva	"
33. Rakahanga	Rierson	"
34. Clipperton*	"	Isolated territory of France
35. Dulcie	"	Great Britain

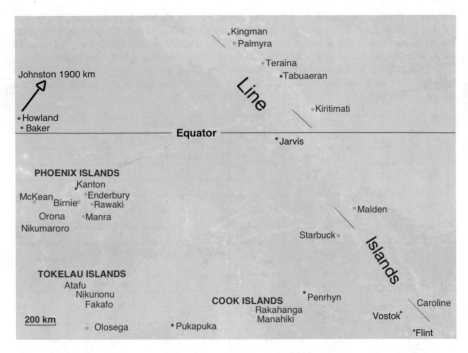

Fig. 6.7 Most of the Phoenix, Tokelau and Line Islands, and the northern Cook Islands shown here in the central Pacific are atolls; Howland, Baker, Jarvis (USA), and Vostok and Flint (Line Islands, Kiribati) are low islands, but not atolls (adapted from maps.fiu.edu/gis/atolls/goldberg)

Manra. Likewise, Malden and Starbuck atolls in the Line Islands were claimed, and all of these were mined although not necessarily by the original claimants. When guano was claimed on Malden Atoll by the U.S. Guano Company, for example, the Company's assertion that the island was U.S. territory was considered null and void by the State Department because it was left unoccupied and un-mined. In the meantime another American company mined Malden, poaching U.S. Guano's claim for 3 years until the island passed into British hands whose leaseholder extracted £100,000 of guano per year there for 30 years [1].

Two of the northernmost Line Islands, Palmyra and Kingman Reef (Fig. 6.7), originally claimed under the Guano Act, remained in American hands. Palmyra became an important naval asset during World War II, but it was never mined for guano. Kingman, typically underwater at high tide, was also never mined and likely did not possess significant quantities of guano. Midway, part of the northwest Hawaiian island chain, and Wake Island were also subsumed as U.S. territory although not specifically through the Guano Act. Secretary of State Seward, the Act's sponsor, was motivated by the need of a U.S. coaling station and a trans-Pacific cable station and that is how the U.S. acquired those outposts. Other islands that were claimed under the Act and remained in U.S. hands included French Frigate

Shoals (part of the Hawaiian Archipelago), and Olosega (Swain's Island), now part of American Samoa (Fig. 6.7).

Johnston Atoll is part of the Pacific Island Remote Area controlled by the United States along with Palmyra and Kingman atolls. At 1400 km NW of Kingman and about the same distance SW of Honolulu, Johnston is quite isolated. Both came under the control of the U.S. Military during World War II and were heavily modified (see Chap. 7), but are now National Wildlife Refuges administered through the U.S. Fish and Wildlife Service. Other remote Pacific islands that came into U.S. possession by way of the Guano Act include Baker, Howland and Jarvis islands (Fig. 6.7), which are also wildlife refuges.

6.3 From Guano to Coconuts

Malden and Starbuck atolls were annexed by Great Britain in 1866, as were most of the Phoenix Islands a few years later. This opened the door for the English entrepreneur John T. Arundel to finish the job of extracting guano on Enderbury, Starbuck and Sydney (the latter now Manra Atoll, but originally named after Arundel's 2nd daughter who was born there). But Arundel's main contribution to these central Pacific islands was to replace native forests with coconut plantations on many of the Phoenix and Line group to feed the boom in copra and palm oil. The oil produced a pure white soap that became a very popular bath item, as well as a source of glycerin for nitroglycerin and dynamite. Arundel planted 2.3 million coconut seedlings from Palmyra to Manra (Fig. 6.7) and even on islands such as Enderbury that supported few trees on its own. Even though most of these plantations failed due to the droughts of 1892–1894, market forces temporarily encouraged other copra plantations to take their place [6]. Even though the coconut palm occurs naturally throughout the tropics, the 'palm fringed shores' advertised in tourist brochures are unlikely to be natural and are much more likely to be plantation remnants [7].

The removal of guano and the disruption of bird colonies on guano islands might have been insufficient by themselves to prevent the birds from recolonizing, but coconut palms were much more effective deterrents. John Arundel and others typically cut down natives such as *Pisonia grandis*, the 'devil's claw tree', which grows to 30 m in coral rock producing multiple widely spread branches (Fig. 6.8a). Coconut palms, by contrast, have branchless trunks and small crowns (Fig. 6.8b) that lack protective nooks and crannies. They are poor places for the support of nests, even for birds that normally inhabit trees. To make matters worse, most guano island birds are colonial and nest in large groups, which coconut palms discourage, and with both the birds and their guano gone, the soil of these once fertile islands probably declined dramatically [8, 9].

Fig. 6.8 (**a**) Mature *Pisonia grandis* trees native to Palmyra Atoll are bird-friendly nesting sites, but have been cut down there and on many other islands and replaced with far less friendly coconut palms (Courtesy of US Fish and Wildlife Service and U.S. Geological Survey) (**b**) Grove of coconut palms on the shore of Palmyra Atoll (Courtesy of Stacie Hathaway, US Geological Survey http://soundwaves.usgs.gov/2011/04/research2.html)

6.4   Guano and the Slave and Coolie Trade

For Peru, where supplies of guano were plentiful, the principal problem was mining it. Extracting guano was not like coal or gold. It did not require shafts or heavy equipment because there it was, on the island surface. All it took was manual labor- and a lot of it- using picks, shovels and wheelbarrows that had to be hauled varying distances to a ship waiting near shore. The initial solution to the labor problem was to bring in workers from Hong Kong, Macau and other Chinese ports. By 1875 anywhere from 90,000 to 150,000 coolies were shipped to Peru, [1] a term derived from the Hindu word 'kuli' meaning 'hired laborer'. The hiring process often included a contract, which the laborer typically could not read. It stipulated that the immigrant would work in a specified region for a certain period (typically 5–8 years) in exchange for the cost of the voyage plus exorbitant expenses for room and board once onsite.

At the end of service the coolie was free to leave. However there were a few hurdles along the way. First, journey from China to Peru took several months and involved accommodations typically used for cargo. There, coolies were typically crowded together and transported in the lightless hold of the vessel just like sacks of flour. As a result the mortality rate due to dysentery, flogging and suicide fluctuated between 20 to more than 30%. On occasion, the coolies fought back as they did aboard the American ship Waverly in 1855. Transporting 450 coolies, some became desperately ill and tried to get on deck but were forced back by the crew. After a month at sea the ship's cooks refused to feed the Chinese unless they received their monthly wages. The coolies panicked and broke through a hatch to reach provisions but some were shot and the crew forced the rest below decks. The next day nearly 300 were found dead, likely by suicide as an alternative to starvation, sickness and a slow death in the dark [10].

The contract, if there was one, also involved chicanery. Some were told that they would be working the goldfields in California. Others were kidnapped outright. Coolies often worked and lived no better than slaves and were given insufficient food and poor medical care while working long hours. The final hurdle included the high probability that a guano worker would not live long enough to see the end of his contract. The Chinchas, bleak volcanic outcrops 30–100 m high, was an unforgiving place. Despite the tropical heat there was virtually no rainfall, and with no trees or shade from the sun the work was unbearable. The dust and ammonia released during excavation caused nosebleeds and eye irritation at first, but after prolonged exposure led to shriveled skin, blindness, lung diseases and gastrointestinal maladies. Given the appropriate wind conditions the stench of guano could sometimes be carried 25 km away where the fumes caused the people of Pisco to cover their faces.

The food for miners included of one cup of maize and four unripe bananas twice a day (which laborers had to pay for), a diet that contributed to the proliferation of scurvy. Those who had the temerity to complain or slow down due to malnourishment were beaten with cat-o'-nine tails or attacked by half-starved dogs. Those who could no longer use their hands were often yoked to wheelbarrows like mules to haul heavy loads to ships waiting at the bottom of cliffs, and in some cases the miners threw themselves over instead of the guano [1, 10].

Fertilizer shipping was only slightly less loathsome an occupation for seamen and shippers. Caustic ammonia fumes, especially from the Chincha Islands guano, permeated the ship and was alleged to overwhelm seamen who were then incapable of proper navigation among other duties. In other instances, coffee and other comestibles that were shipped in the same hold became contaminated with the odor and were rendered worthless. Legal action was taken against the ship owners to recover damages. A ship that transported guano either had to be fumigated to rid itself of the stench or it became consigned to guano transport only. Even then, anything made of iron was immediately attacked by the fumes and began to rust. That included anchors, cables, and bolts that held the hull timbers together. The maintenance costs were steep relative to the profit expected by the shipping companies [11]. Something had to be done.

The coolie trade soon became criticized as a new form of slavery and in 1856 Peru made it illegal. However, the old-fashioned form of slavery (no contract and no pay) was still in effect. For the first few years after 1856, Melanesians were 'blackbirded' into the Peruvian guano trade, but by the 1860s the practice included Polynesians, especially those who were much closer to Peru's home ports. On some occasions, the native people went willingly, as in the case of 250 residents of Penrhyn Atoll, in the Cook Islands who were suffering from food shortages and famine-induced warfare. On other occasions, however, islanders were not so eager to leave their homes. In June 1863 whaling Captain Thomas James McGrath invited about 350 people from Tonga on board for trading. But once almost half of the population was on board, the ship's doors and rooms were locked, and the ship sailed away. Similarly, a score of slave ships set out from Callao and abducted nearly the entire adult male population from Easter Island to work the guano mines. Ironically, after slavery was made illegal in Peru, 470 Easter Islanders were hastily crammed into a Peruvian ship for repatriation but 162 of them died before the ship left port due to an outbreak of smallpox. Only 15 lived long enough to see their homeland and it is unclear how many died from the disease the survivors brought back with them [6].

The United States outlawed slavery in 1863, but employed similar tactics in employing laborers. This is no better illustrated than by a case that came before the Supreme Court in 1890. The Court had several questions before it including the constitutionality of the Guano Islands Act, and the jurisdiction of American courts on the Caribbean island of Navassa, which was claimed by the Navassa Phosphate Company under the Act. Henry Jones was one of several black defendants who was tried and convicted of murder on the island. Testimony at trial revealed that corporal punishment was meted out routinely for trivial offences, that labor was both hard and dangerous, that the food was extremely poor, and the men were gouged and indebted to the company store. They claimed that if they were allowed to leave they would have to forfeit all of their earnings, but they often were not allowed to leave before their contract was completed, and sometimes not even afterwards. Anger boiled over to conspiracy, as some witnesses put it. Others suggested it was a spontaneous uprising. Four supervisory (white) employees were killed and in the in the trials and appeals that followed, fourteen black laborers at Navassa were convicted of manslaughter and sentenced for two to ten years at hard labor. Mr. Jones and two other laborers were found guilty of murder and were sentenced to be hanged. In short, the court stated that acquisition of territory was a political rather than a judicial question, and that Congress was within its constitutional bounds when it authorized the Act. Moreover they found the extension of criminal jurisdiction over territory taken under the Act was simply an extension of admiralty law. Offenses or crimes committed on any such island or in the waters adjacent thereto, 'shall be held and deemed to have been done or committed on the high seas, on board a merchant ship or vessel belonging to the United States, and be punished according to the laws of the United States'. The court upheld the convictions, but the three condemned men applied for executive clemency. President WH Harrison in reviewing the facts agreed that the Company was guilty of egregious behavior as a mitigating set of circumstances, and commuted the sentences of Jones and the other two men to life in prison [12].

6.5   The Nitrogen Wars and the End of the Guano Trade

Despite the control of the world's fertilizer industry for decades, several factors caused the Peruvian economy to come to grief. First, Peru had come to rely almost entirely on guano income, and for a variety of reasons, little of it was used for diversification and reinvestment. Secondly, by 1876 there was little guano left to extract, and at about the same time, the price began to plummet (Fig. 6.9) from competition by superphosphate, as described below. This caused Peru to default on its loans to Great Britain, which at the time amounted to more than $32 million in U.S. dollars [13]. Then in 1879, just to put another nail in its coffin, Peru became embroiled in a tax and border dispute with Chile that culminated in a four-year conflict, the War of the Pacific, sometimes called the Nitrogen Wars or (inappropriately) the Guano Wars.

In 1879 the borders of Peru, Bolivia and Chile were considerably different from their present state. Peru's southern border extended south into the Atacama Desert, the driest non-polar desert in the world, a region that is now part of Chile. Likewise, the state of Bolivia, now landlocked, once extended west to the Pacific coast, where it occupied the Atacama region south of Peru (Fig. 6.10). There, all three nations were engaged in the exploitation of nitrate of soda (sodium nitrate) that occurred in enormous deposits just beneath the desert surface. Nitrates could not only be used in fertilizer production, but could also be employed in the production of sulfuric and nitric acids, both of which had widespread applications to glass, glaze, and bleach

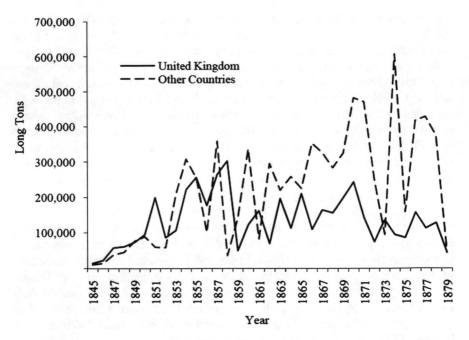

Fig. 6.9 Peruvian guano exports to The United Kingdom and other countries plummeted in the early 1870s and after a brief resurgence came to a virtual end in 1879 (Courtesy of Catalina Vizcarra, University of Vermont [4])

production among many other uses. And with a little chemical tinkering, sodium nitrate from the Atacama could also be used to make explosives such as gunpowder and nitroglycerin, both of which were in high demand for mining and warfare. The problem with nitroglycerin was its tendency to explode unexpectedly, but Alfred Nobel (after whom the Peace Prize is named) invented a stabilized form of it that he called 'Dynamite', the use of which soon rendered gunpowder and nitroglycerin explosives obsolete.

By 1878–79 Peru was suffering not only from economic downturn due to low guano production and prices, but along with Bolivia experienced a series of natural disasters including a strong earthquake and a tsunami, as well as a rare flood in the Atacama from the strongest El Niño of the nineteenth century.

El Niño is a periodic climate event that occurs every 4–7 years (Chap. 1) when winds that force upwelling become anemic or cease entirely. Its effects in the Pacific islands have been described already, but the effects in Peru are different. El Niño wind shifts allow warm tropical water to intrude along the Peruvian coast, a region that is usually under the influence of the cool Humboldt Current. The interruption of the cold-water flow causes a decline upwelling, in plankton production, and a dramatic regression of the anchoveta population. Not surprisingly, these changes also have a devastating effect on the seabird population. Not only is their food supply disrupted, but the increase in heat and humidity causes adult pairs of birds to abandon their nests and their chicks. Strong El Niños are often accompanied by substantial

Fig. 6.10 Borders of Perú, Bolivia and Chile before and after the Nitrogen Wars. The southern border of Peru (*hatched red*) and the southwestern coastal border of Bolivia (*hatched green*) are depicted. After 1884 both hatched areas became Chilean territory and remain so today (Map courtesy of Wikipedia.org)

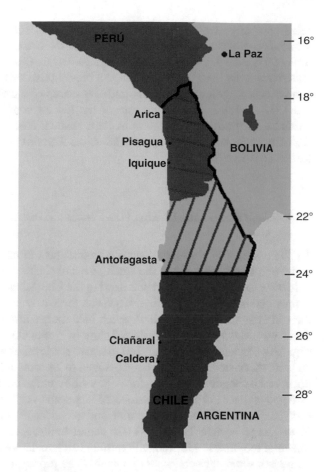

rainfall that can wash out nesting sites, and the adults themselves often succumb to disease from the stress if they lack the energy to migrate. Add to this poaching of bird eggs and wholesale destruction of ground-nesting colonies by guano miners who found the birds 'in the way', or used them for food, or took their feathers to be used as a source for the fashion trade, and the result was the loss of the source of guano itself. While these events were occurring along the coast, the breadbasket of Bolivia and Peru, farms in the highlands, were experiencing the opposite effect of El Niño, a severe drought that caused massive crop failures. Collectively, these were the final blows to the Guano Age [6].

In light of the resulting decline in tax revenues from both guano and food production, Bolivia decided to impose a modest tax of 10 centavos per metric ton on nitrate exports in its coastal territory. However, the largest of these exporters was a British-Chilean consortium called the Antofagasta Company and they refused to pay. Antofagasta reasoned that the tax violated a treaty signed several years before, that encouraged such companies to set up shop in Bolivia tax-free for 25 years. Undeterred by all of this treaty business, the Bolivian government then began to seize company assets to cover unpaid taxes, and that is when Chilean marines invaded coastal Bolivia. Chile was aware that Bolivia had signed a mutual defense treaty with Peru and asked the Peruvians to declare their neutrality. When no affirmative response came, Chile launched a pre-emptive strike to the north into Peru's Atacama, took their guano islands, and even occupied the city of Lima. The Nitrogen Wars (1879–1884) had begun, one of the largest conflicts ever fought in Latin America. A protracted civil war in Peru ensued. Recovery of the Peruvian and Bolivian economies would take decades, the Age of Guano was over, and Bolivia lost its coastline [6].

6.6   Superphosphate and the Pacific Islands

By the end of the nineteenth century, extraction of guano on Pacific islands had come to a virtual halt due to the small size and shallow extent of the deposits they held. One exception was Starbuck Atoll in the Line Islands (Fig. 6.7), whose guano supply persisted until 1927 [14]. However, three new island sources were found in the early twentieth century, all of which held more substantial deposits. The first of these was Makatea in the western Tuamotu Archipelago (French Polynesia, see Fig. 5.4), elevated to more than 100 m above sea level. It was mined until its deposits were exhausted in 1966. There appeared to be some relationship between island uplift and a chemical change in the composition of its limestone that did not involve bird poo- at least not directly. Instead, Makatea and similar islands exist in relatively rainy environments where bacterial and physical degradation of guano slowly release its phosphate. This allows it to percolate through and react with the underlying coral reef limestone, gradually turning calcium carbonate into a high-phosphate carbonate sometimes called 'rock guano'. The process of its formation is complex and involves phosphate incorporation into the rock structure over long periods of time, perhaps from alternating periods of exposure and immersion by changes in sea level. Alternatively, rock guano may have formed in isolated lagoons where phosphate concentrations in the water were enhanced [15, 16]. In either case, these deposits may contain up to four times the amount of that mineral compared with guano itself, and in the fertilizer world that rock is sometimes referred to as 'superphosphate'. When combined with cheap Bolivian or Peruvian nitrate of soda, or potassium from Europe, early twentieth century fertilizer firms were able to custom formulate plant foods.

A similarly uplifted volcanic island called Banaba near the Gilbert chain (Fig. 6.11) was discovered with enormous amounts of high-grade rock phosphate deposits on its surface. An Australian engineering firm estimated the reserves at 13 million tons, an amount that was larger than the gross weight of all of the guano taken from Peru during its golden age. When its geological value was discovered there were 451 native Banabians who called this island home. But because no European powers had claimed it, the Royal Navy promptly raised the flag over Banaba and made the island part of its Gilbert and Ellice Islands Protectorate. John

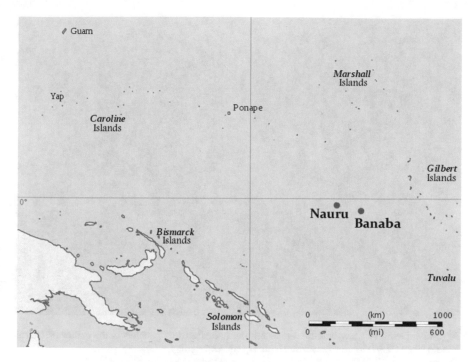

Fig. 6.11 Location of the isolated phosphate islands of Banaba and Nauru (Courtesy of Wikimedia. org)

Arundel's company swiftly contracted with the island's chieftain, giving it the sole right to mine and ship all of the phosphate on the island for 999 years in exchange for £50 per year [6]. It did not take long for Banabians to realize that this deal was grossly lopsided, but strip mining continued and by 1981, 17.7 million metric tons of high-grade phosphate had been taken through colonial authority. That included mining and destroying the best food-producing area on the island's central plateau. Banaba was not the only place where this happened.

Nauru Island, a Cautionary Tale

Nearby isolated Nauru, another 20 km^2 dot up to 60 m high (Fig. 6.12) was discovered and was estimated to contain more than three times the amount of phosphate rock on Banaba. Interestingly, Germany was expanding its claims in the Pacific as was England, and the two countries decided to divide the western Pacific along certain lines of demarcation in 1886. Great Britain colonized the Gilbert and Ellice Islands, and claimed the British Solomon Islands as well. The demark fell immediately between Banaba on the British side, and Nauru on the German side [17]. Germany had already claimed part of Papua New Guinea along with the Bismarck

Fig. 6.12 Detail of phosphate-limestone mining operation that pockmarks the center of Nauru (Photo taken (2007) by Lorrie Graham courtesy of Wikimedia Commons)

Archipelago and part of the Solomon Islands, and they annexed Nauru two years after the agreement. The growing extent of German acquisitions in the Pacific would set the scene for an expanded sphere of influence prior to World War I as described in Chap. 7.

Germany extracted phosphate on Nauru after its discovery in 1900, but after World War I it became British territory and was governed by the British Phosphate Commission operated by the United Kingdom, Australia and New Zealand. The change in ownership came with a demand for miners that exceeded the modest supply on both Banaba and Nauru. As a result company ships brought in hundreds of workers, mostly Ellice and Gilbert islanders, followed by those from Japan, China and the United Kingdom. Poor working conditions and food rationing used to punish complaining workers promoted ill health. Human waste disposal quickly became a problem, followed by a dysentery epidemic. Then there were outbreaks of tuberculosis, whooping cough and pneumonia, likely abetted by the ubiquitous phosphate dust. There were also repeated polio epidemics [6].

Prior to mining there was a culture of fishing and gardening among the Nauruans who ate fresh seafood, as well as fruits and vegetables grown on land. However, strip mining that went on for decades by various national consortia caused the natural landscape to disappear. As much as 80% of the island is now barren, and pock marked by limestone pinnacles, some more than 20 m tall (Fig. 6.12). The mining has even changed the weather. Heat that rises from the mined-out plateau drives away rain clouds, leaving the island plagued by long periods of drought. The only remaining inhabitable land is a narrow coastal fringe (Fig. 6.13). Mining not only

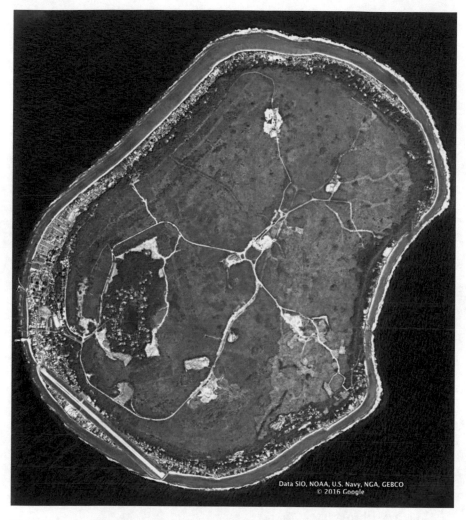

Fig. 6.13 Isolated Nauru is a flat-topped equatorial atoll raised to 60 m above sea level. Most of the island has been heavily strip-mined as is evident from the access roads that cross the treeless central portion. Almost all of this island nation's people live in a narrow margin along the coastline. An airstrip, 2100 m long can be seen on the southwest side (Image: Google Earth)

affected the terrestrial realm, it also contaminated the marine environment with silt and metals after rain carried the mine tailings into the ocean. Nearshore reefs and fisheries in many areas have been affected [18]. Food is now primarily imported from Australia with a great reliance on white rice, soda, instant noodles and 'anything in a tin' [19]. As a result, Nauru has one of the highest rates of obesity and diabetes in the world. Fully 94% of its residents are overweight while 72% of them are obese. More than 40% of the population has type-2 diabetes, as well as other significant dietary-related problems such as kidney and heart disease.

When the island was granted independence in 1968 it began attracting outside investors who would pay market value for its phosphate reserves. A trust fund was created, the income from which allowed the island's 7500–10,000 residents to boast of the highest per-capita income of any sovereign state in the world. However, the funds were mismanaged and poorly invested. By the 1990s most of the phosphate rock had been depleted and when exports began to slow, the island began looking to other ventures to supplement its income. For a few years the island became a haven for offshore banking, or more accurately, a center for tax avoidance. For as little as $25,000, anyone could set up a bank account in Nauru without ever setting foot on the island. Other tax havens require banks to record transactions while keeping the details from the prying eyes of foreign officials. However, banks incorporated in Nauru were not required to keep records at all [20]. In 2001 the licenses of hundreds of Nauru banks were revoked due to an international blacklist. A few years later The Bank of Nauru itself closed and since then the island has operated on a cash economy; savings were buried in the back yard [21]. Nauru still exports small amounts of phosphate but the revenue earned is not enough to sustain the population. Foreign aid from Australia is a major source of the island's income.

References

1. Skaggs JM (1994) The great guano rush. Entrepreneurs and American overseas expansion. St Martin's Griffin, New York
2. Davidson EA, Howarth RW (2007) Nutrients in synergy. Nature 449:1000–1001
3. Krebs C (2008) The ecological world view. University of California Press, Berkeley
4. Vizcarra C (2009) Guano, credible commitments, and sovereign debt repayment in 19th century Peru. J Econ Hist 69:358–387
5. Moore JB, Wharton F (1906) A Digest of International Law of the United States 1:556–580. A digest of international law: as embodied in diplomatic discussions, treaties and other international agreements, international awards, the decisions of municipal courts, and the writings of jurists …
6. Cushman GT (2013) Guano and the opening of the Pacific world. A global ecological history. Cambridge University Press, Cambridge
7. Mueller-Dombois D, Fosberg FR (1998) Vegetation of the tropical Pacific Islands. Springer, New York
8. Young HS, McCauley DJ et al (2010) Plants cause ecosystem nutrient depletion via the interruption of bird-derived spatial subsidies. Proc Natl Acad Sci USA 107:2072–2077
9. Smith JL, Muldur CPH, Ellis JC (2011) Seabirds as ecosystem engineers: nutrient inputs and physical disturbance. In: Muldur CPH, Anderson WB et al (eds) Seabird Islands, Oxford University Press, Oxford
10. Schultz A (2011) American merchants and the Chinese coolie trade 1850–1880: contrasting models of human trafficking to Peru and the United States. Western Oregon University. http://digitalcommons.wou.edu/cgi/viewcontent.cgi?article=1006&context=his
11. Hollet D (2008) More precious than gold: the story of the Peruvian guano trade. Fairleigh Dickinson University Press, Teaneck
12. Jones V. United States U.S. 202 (Nov. 24, 1890): error to the circuit court of the United States for the district of Maryland. https://supreme.justia.com/cases/federal/us/137/202/case.html

13. The great Peruvian guano bonanza, rise, fall and legacy. Council on Hemispheric Affairs. http://www.coha.org/the-great-peruvian-guano-bonanza-rise-fall-and-legacy/
14. Hein JR (2009) Phosphate islands. In: Gillespie RG, Clague DA (eds) Encyclopedia of Islands. University of California Press, Berkeley, pp 738–740
15. Rossfelder AM (1990) The submerged phosphate deposits on Mataiva Atoll, French Polynesia. In: Burnett WC, Riggs SR (eds) Phosphate deposits of the world 3. Cambridge University Press, Cambridge, pp 195–204
16. Piper DZ, Loebner B, Aharon P (1990) Physical and chemical properties of the phosphate deposits on Nauru, western Pacific Ocean. In: Burnett WC, Riggs SR (eds) Phosphate deposits of the world 3. Burnett Cambridge University Press, Cambridge, pp 177–194
17. Fortune K (2000) In: Lal BV, Fortune K (eds) The Pacific Islands: an Encyclopedia. University of Hawai'i Press, Honolulu, p 212
18. Sulu R (2007) Status of coral reefs in the southwest Pacific 2004. IPS Publications, University of the South Pacific, Fiji, pp 83–94
19. Hallet V (2105) The people of Nauru want to get healthy- so why can't they succeed? National Public Radio. http://www.npr.org/sections/goatsandsoda/2015/09/22/442545313/the-people-of-nauru-want-to-get-healthy-so-why-cant-they-succeed
20. The Economist December 20, 2001. Paradise well and truly lost. http://www.economist.com/node/884045
21. Stewart RM, Craymer L (2013) Nation of Nauru rebuilds bank system from scratch. Wall Street Journal November 3, 2013. http://www.wsj.com/articles/SB10001424052702304200804579165092731005968

Chapter 7
Domination of Pacific Islands in War and in the Nuclear Age

Abstract While the claiming of Pacific islands for Europe began early with the Spanish explorers, by the end of the nineteenth century almost all of them belonged to one power or another. German expansion began in 1884 through claims or purchases and was followed in kind by those of the British and the United States. Prior to World War I the German flag was flying over much of the western Pacific, but this changed dramatically at the end of the war when Japan assumed control of the German islands and promptly began to fortify and otherwise militarize them. The use of Pacific islands during World War II is briefly reviewed with emphasis on their alteration in preparation for war. The postwar period focuses on the use of Pacific islands for nuclear testing by the United States, Britain and France, and the movement toward decolonization and independence. However, for many island groups that had been militarized, the disruption that came with it did not go away with declarations of independence. The alteration of Pacific island culture, and the culture of dependence brought about during the twentieth century are described.

7.1 Changes Before and After WWI

By the late 1800s most of the islands in the tropical Pacific had become colonial territories of the European powers. Germany made an aggressive territorial entrance to the Pacific after achieving its own unification in 1871. Many in Germany at the time believed that true nationhood required the acquisition of overseas colonies. German chancellor Bismarck was not necessarily a proponent of such outposts, but he became convinced that annexations would best be handled by granting charters to private companies backed by imperial protection. Thus when a syndicate of German bankers founded the New Guinea Company for the purpose of colonizing and exploiting resources in the region, a charter was granted in 1884 and Germany promptly took possession of northeastern New Guinea, renaming it Kaiser Wilhelmsland. The Bismarck Archipelago was annexed in the same year, and was followed shortly thereafter by the Marshall Islands and the northern Solomon Islands as protectorates [1]. Thus began the German colonial empire in the Pacific. In response, Great Britain declared as protectorates the southeastern quadrant of New Guinea as well as the southern Solomon Islands. These divisions led to the

© Springer International Publishing AG 2018
W.M. Goldberg, *The Geography, Nature and History of the Tropical Pacific and its Islands*, World Regional Geography Book Series,
https://doi.org/10.1007/978-3-319-69532-7_7

Table 7.1 The division of selected Pacific Islands by Europe & the U.S. Before WWI [1, 4]

Island/ Island Group	Protectorate of/Annexed by
Bismarck Islands	Germany, 1884
Caroline Islands	Germany agrees to annexation by Spain, 1885; Germany, purchases islands from Spain, 1899
Fiji	British Crown Colony, 1874
Gilbert & Ellis Islands	Britain, 1886/1892
Hawaii	United States, 1898
Johnston Atoll	Claimed under the U.S. Guano Act, 1858
Mariana Islands	Germany, purchased from Spain, 1899 with the exception of Guam, taken by U.S. after the Spanish-American War in 1898
Marshall Islands	Spain cedes islands to Germany1885, in exchange for Caroline Islands and compensation
Midway Atoll	Claimed under the U.S. Guano Act, 1867
Northeastern New Guinea Southeastern New Guinea Western New Guinea	Germany 1884 Britain 1884/1888 The Netherlands, 1828
Nauru Island	Germany, 1888
Palmyra Atoll	Claimed under the U.S. Guano Act, 1859
Samoa western islands	Germany, 1899
Samoa eastern islands	United States, 1899
Solomon Islands	Northern islands declared a protectorate by Germany, 1886; 1885; Southern islands by Britain, 1893
Wake Atoll	Formerly claimed by Spain, Annexed by United States after Spanish-American War in 1899

Anglo-German Demarcation Declaration of 1886, which ensured that the British would also hold territory in the Gilbert and Ellis groups, in exchange for German control elsewhere as described below. Even so, there were territorial disputes and claims during this period of time, as summarized by Table 7.1.

Several islands in Samoa that were not included in the Declaration became German Samoa as a result of business interests, primarily in coconut, cacao, rubber and cocoa cultivation. However, German enterprise conflicted with American and British concerns, which led to civil war and military confrontations among the three parties. A tripartite agreement in 1899 led to dividing Samoa into German territory in the west, while islands in the east were designated as American Samoa. England for its part did not get a piece of Samoa, but she took over German interests in Tonga (although the Kingdom of Tonga itself remained independent), as well as the formerly German islands in the Bismarck and Solomon archipelagoes [2]. It does not appear that the native people were consulted on any of these trade agreements. In some cases there were no agreements at all. This was certainly the case on the tiny island of Nauru on the German side of the demarcation line where German marines rounded up all 12 chiefs and held them hostage until they turned over their weapons. With this diplomatic maneuver firmly implemented, the German flag was raised the next day [3]. Phosphate deposits were discovered there in the following year and its commercial extraction began shortly afterward (Chap. 6).

The Mariana Islands and the Caroline Islands had been claimed by Spain since the sixteenth century, but their assertion of dominion was challenged by nineteenth century Germany. A trade was arranged in 1884 and Spain was 'allowed' to take possession of the Caroline Islands in exchange for the Marshall group. However, because of financial issues that resulted from losing the Spanish-American War, the Caroline and Mariana Islands were sold to Germany in 1889. All of these considerable holdings from New Guinea to the Marshall Islands constituted the Imperial German Pacific Protectorates and they were acquired within a decade. Guam, the largest island in the Marianas, was an exception as it became American territory after the Spanish-American War, as did lonely, but strategically significant Wake Atoll (Fig. 7.1).

Germany's Pacific possessions would change dramatically during World War I. In the first week of the war, Japan proposed to enter against Germany if it could take over their Pacific territory. And in 1914 the Japanese Imperial Navy did exactly that, taking over Micronesia including the Caroline, Mariana and the Marshall Islands with virtually no resistance [5]. These strategically important island groups together occupied an area nearly equal to that of the continental United States, even if the total land area was less than 5% of the state of Rhode Island. At roughly the same time Australian forces captured Kaiser Wilhelmsland, as well as the larger islands in the Bismarck Archipelago. After the end of the war in 1919, the League of Nations mandated that all of the German Pacific Protectorates were to be given control by the allies. The Caroline, Mariana and Marshall Islands were to be administered by Japan, and the Australians would take over the German portions of New Guinea, the Bismarck and Solomon islands.

7.2   The Militarization of the Pacific Islands

Under the terms of the League of Nations Mandate, Japan was to demilitarize its islands. However, following the initial Japanese occupation, a policy of secrecy was adopted and by the late 1920s Tokyo was rejecting requests for international inspection or even entry of ships from its wartime allies. During the 1930s, the Imperial Japanese Navy began construction of airfields, fortifications, ports, and other military projects in the islands, viewing them as critical in the defense of the Japanese homeland against a potential US invasion [6]. These islands became important staging grounds for later Japanese air and naval offensives in the Pacific war. For example, Kwajalein Atoll (see Fig. 7.7) in the Marshall Islands became a major base for the support of the December 1941 attack on Pearl Harbor and the strike on Wake Island, an atoll 3700 km west of Honolulu that was initiated at about the same time. Three days later Japanese naval forces also used Truk (now Chuuk) in the Caroline Islands to attack and occupy the Gilbert Islands. Tarawa Atoll in the southern Gilberts became especially well fortified. Likewise, Palau in the Carolines was used to support the invasion of Guam and the Philippines. A few months later Japan invaded the Australian mandate of the Solomon group including the island of Guadalcanal.

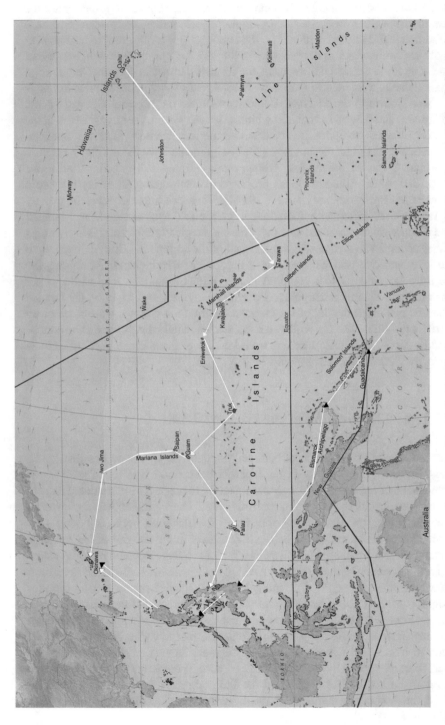

Fig. 7.1 Maximum extent of Japanese occupation is outlined in red. Pacific Islands were re-taken beginning with Guadalcanal in 1942 and a later push toward the Philippines and Okinawa (*dark arrowheads*). A second line of offense began more than a year later at Tarawa Atoll, moving toward Okinawa (*white arrowheads*). Place names of major island battles and other place names used in this chapter are labeled in dark font

The distance of these islands from home complicated fortification by the Japanese, and although concrete and steel were used in defending important positions, local materials and local slave labor were typically required. Coral rock could be pounded into powder and mixed with seawater to produce cement. When mixed with small limestone rock, the cement could be strengthened to form a concrete. Coconut trees were also valued in construction, and because the trunks were extremely resilient due to their soft and fibrous interior, they were widely used for fortifications. Mortar and artillery shells often bounced after hitting them and failed to detonate. Coconut log walls proved to be impenetrable by amphibious vehicles. Logs also were used to support gun emplacements that were otherwise covered with dirt or rock, or they could form double walls filled with sand to form a bunker. Palm fronds were used for camouflage of such positions [7].

Tarawa was fairly typical in the use this type of construction, especially on one of the islands of the arrowhead-shaped atoll. This was Betio and it was here on this 4400 m long and 700 m-wide sliver of land that the Japanese built an airfield considered a threat to shipping and communication between Pearl Harbor, Samoa, Australia and New Zealand. For the United States, Betio was also strategically located as a launch point to attack the Marshall Islands (Figs. 7.1 and 7.2). Small though it may have been, this island was crisscrossed with defenses including 100 concrete bunkers, seawalls, and an extensive trench system for defensive troop movements. Coastal and antiaircraft guns, heavy and light machine guns, and light tanks supported the airstrip. Betio's beaches and shallow reefs were covered with barbed wire and mines. Its forest was essentially stripped for construction, and naval gunfire, bombs and napalm would obliterate what was left (Fig. 7.2). During the 76-hour battle on this 100-hectare island in November 1943, more than 1000 US Marines were killed and 2200 more were wounded. Nearly all of the 4500 Japanese defenders were also killed or committed suicide. There are essentially no accounts of the flora and fauna on Betio before the attack. Much of it could have recovered, but Tarawa changed from a sleepy island with 400 inhabitants who were there prior to the Japanese arrival, as will be described in Chap. 8. A better, albeit limited view of ecological changes wrought by World War II can be seen on other islands, especially three that were held by the United States: Palmyra, Johnston and Midway atolls.

7.3 Case Examples of Island Alteration: Midway, Palmyra and Johnston Atolls

The war in the Pacific had enormous effects on island biota. On both steep volcanic islands and coral atolls, conflict produced levels of ecological degradation of forests, watersheds, coastal swamplands, and coral reefs that had no previous parallel. Atolls in particular are low islands with limited habitat in shallow water and especially on land. They have nutrient-deficient soils and support limited numbers of

Fig. 7.2 Betio Island at the southern end of Tarawa Atoll in the Gilbert Islands as Japanese airfield was being prepared in September 1943. After the fighting and heavy losses on both sides, the island had been pummeled, bombed and stripped of virtually all vegetation. *Inset*: Aerial view of Betio, Tarawa Atoll after the November 1943 assault as seen from a Douglas SBD-5 Dauntless aircraft flying at an altitude of 150 m (Courtesy of U.S. Navy National Museum of Naval Aviation and Wikipedia.org, Battle of Tarawa)

plants and animal species. Both terrestrial and marine, environments are exceptionally vulnerable to human disruption.

Midway Atoll is part of the northwest chain of Hawaiian Islands. Originally, it would have been best described as an atoll reef ecosystem (Chap. 1) composed primarily of broad sandy reef flats that grade into an apron surrounding a small lagoon. Island development was minimal, but modification of the atoll started early. A channel was blasted through to the lagoon in 1871 and in 1935 Pan American

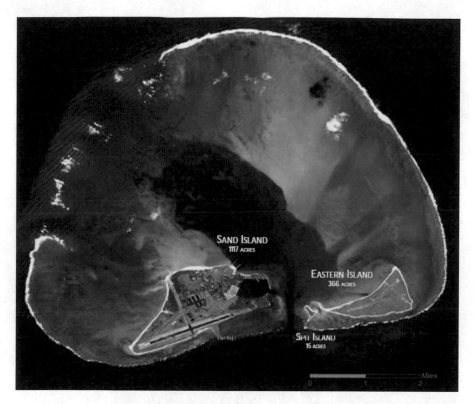

Fig. 7.3 Satellite image of Midway Atoll including its islands and surrounding shallow reefs (Courtesy of U.S. National Fish and Wildlife Service)

Airways established a seaplane base there along with support buildings. The airline brought in 100 tons of topsoil to Midway from Guam so that non-native trees and shrubs could be planted. Streets, piers, a Y-shaped hotel with 45 rooms, and a swimming pool were also built for their overnight visitors [8]. The major developments began with the enlargement of the channel and construction of a submarine and seaplane base by the U.S. Navy in 1938. Construction proceeded more quickly shortly thereafter due to the threat of war. The land area consists of two main islands (Fig. 7.3) that constitute slightly more than 600 hectares or about 8% of the shallow atoll surface (measured using Google Earth Pro). These dredged islands were flattened, paved for the construction of roads and airfields, and filled for the support of storage buildings and barracks. This work went on for 18 months, day and night. Tropical forest habitat was leveled. Ground- and burrow-nesting birds such as several types of terns, albatrosses, shearwaters and petrels were not only displaced during construction, but were crushed along with their eggs by bulldozers and other heavy equipment. Large and numerous albatrosses were especially problematic during the nesting season (Fig. 7.4) due to their tendency to gather on runways where they were killed every time a plane landed or took off. Black rats were introduced

Fig. 7.4 Laysan albatrosses nesting at Sand Island, Midway, March 2015 (Courtesy of Forest and Kim Starr, USFWS)

to the island in 1943 and exterminated the Laysan finch and the flightless Laysan rail. Invasive plants and animals also became pests [9, 10].

Midway was attacked in mid-1942, but its significance (the Battle of Midway) was as an air and naval battle north of the atoll in which Japan lost a substantial part of the Imperial Navy, especially its aircraft carriers. American cryptographers had broken Japan's code and knew in advance where they were concentrating her fleet. It was a major turning point of the war.

Johnston Atoll was one of many claimed under the Guano Islands Act (Chap. 6). However, its ownership was disputed by Hawaii, a point that became moot when Hawaii was admitted as a U.S. territory and Johnston was declared a bird sanctuary. At 1400 km SW of Honolulu, Johnston is quite isolated, and that led to its military value. Two islands on the atoll were natural and comprised 23 hectares, but in 1936 the Navy began the first of many changes to enlarge the atoll's land area. Coral rock on a small sand cay was blasted to clear a runway that would accommodate seaplanes flying from Hawaii. Johnston Island was the larger of the two and was originally less than 20 hectares, but it was eventually expanded more than tenfold and became dominated by a 1200 m long runway (Fig. 7.5). A dredge channel was constructed around the island accomodate ship movement. Two additional islands were artificial creations of dredge spoil, making the total expansion of land area 12 times the original.

Fig. 7.5 Johnston Island, the main part of the atoll in 2016. Note main airstrip at center and dredge channel extending around the right half of the island. Courtesy of DigitalGlobe. Inset: Enlargement sequence of Johnston Island from 1942–1964 (Drawing by S. Skartvedt, courtesy of Wikipedia.org)

The island became the site of high-altitude nuclear tests after the war from 1958 to 1963. Some launch failures caused serious radioactive contamination to the islands and its lagoon that remain an issue to this day. Since the 1960s, the islands on Johnston Atoll have been used as an anti-satellite missile base, a satellite tracking station, and as a site for chemical waste storage. An incinerator-based disposal system for Agent Orange, Sarin, VX, mustard gas, and other munitions was established in 1985 and was active until 2002. Despite once being offered for sale by the U.S. military as a vacation spot in the late 1980s, the island remains heavily contaminated and is an (unfunded) Environmental Protection Agency Superfund site. Ironically, it has once again become a bird sanctuary managed by the U.S. Fish and Wildlife Service [11, 12], although entry requires a special permit from the U.S. Air Force.

Palmyra Atoll is located in the Line Islands, about 1600 km southwest of Honolulu where it occupies a surface area of 12 km². The changes to this island's ecosystems are well documented. The pre-war atoll had three large lagoons (western, eastern and central) that were nearly surrounded by shallow reef flats. A scattering of small islands formed a chain adjacent to the reef flat, but the entire lagoon system was open to incoming tides, especially on the western side of Palmyra where there were no islands. The atoll was initially modified as a coconut plantation. Two hundred coconut palms were planted in 1885, but by 1913 it was estimated that there were at least 25,000 [13]. By 1940 several islands had changed shape and some were joined together, but the largest changes were just beginning. A 60 m-wide ship channel was dredged and blasted, connecting the westernmost lagoon to deeper water. The dredge spoils created several new islands and greatly enlarged others, some large enough for airstrips. Another supported 25 buildings. By the time the dredging was complete, the central and western lagoons were joined to create a turning basin, and a roadway isolated the eastern lagoon. The roads were 1–2 m high and connected all of the islands except those on the western part of the rim, effectively cutting off tidal circulation to the lagoon [13, 14] as shown in Fig. 7.6). At the height of the war there were 6000 military personnel on Palmyra and all of the attendant supply and sewage disposal problems that accompany such numbers. While the disruption to the terrestrial environment was clear enough, a postwar examination of the reefs outside of the rim suggests that they were and still are nearly pristine [13]. By contrast, the damage to the lagoon system was somewhat subtler.

Before construction, large waves likely pushed relatively cool ocean water across the lagoon system from the northeast to the west, and moved it successively into each lagoon. The cooler waters would have been relatively dense and would have sunk to the bottom renewing the deeper waters with oxygen. However the closing of the natural reef channels impeded tidal exchange across the reef flat. This undoubtedly destroyed the natural circulation of the water in the lagoons, and perhaps caused an increase in water temperature. In addition, the dredged channel allowed strong tidal currents to enter the west lagoon, which stirred up fine sediments that had lain undisturbed on its floor for thousands of years [15]. These alterations killed much of the lagoon's marine life, smothering them in a shroud of fine sediment. However, there are indications that the modified islands and the causeways are gradually eroding, albeit with a concomitant redistribution of sediment [16]. While there is a dispute concerning how best to promote what is left of the lagoon's reefs [17], they show little sign of recovery.

After construction these lagoons not only lacked corals and other reef organisms of any significance, but they also gave off a strong odor of toxic hydrogen sulfide (familiar as the smell from rotting eggs) [13]. There has been some recovery in shallow water, but the lagoons are up to 55 m deep and below 25 m the water lies undisturbed, lacking any renewal by regularly circulating water. Correspondingly, below that depth the lagoons are now oxygen-deficient [14] as indicated by Fig. 7.6.

7.4 American Nuclear Tests in the Marshall Islands

In 1947 the United Nations formally revoked Japan's League of Nations mandate of Micronesia, making the United States responsible for the Caroline, Mariana and Marshall Islands in the form of a trusteeship. The administration of the Trust Territory was unusual. Rather than answering to the general assembly,

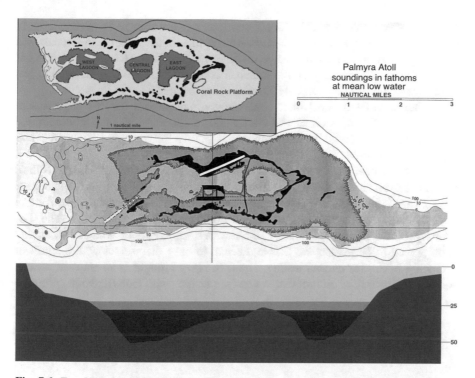

Fig. 7.6 *Top:* Map drawn in 1913 showing separate lagoons (*dark blue*) and separate, small islands (*black*). Shallow reef and reef flats are shown in light *green*. Colored after black and white map by Rock, 1916 in Dawson 1959 [13]. *Center:* Palmyra Atoll postwar construction, NOAA chart from 1991. Note ship channel lower *left*, and causeway isolating the east lagoon from the western and central lagoons (now joined). The runway on Cooper Island (*white* strip at north) is 1.6 km long, and is the longest of four made from crushed coral rock. The lagoons are surrounded and bisected by interconnecting, elevated roadways except on the western side of the atoll. NOAA map courtesy of the University of Texas map collection. *Bottom:* Cross-section of lagoons showing oxygenated water (*blue*), transition layer (*orange*) and anoxic hydrogen sulfide layer (*red*) filling the lagoons below 25 m. Depth at right in meters. Re-drawn from Gardner et al. [14]

administrators only had to bring matters to a vote with the Security Council where the U.S. had a veto. That made the Trust Territory in essence an American colony, but one that would be difficult to manage. There were more than 100,000 people scattered over an area nearly the size of the continental United States who were members of at least six ethnic groups. There were also nine mutually unintelligible languages, even within the same island chain. For example in the Caroline group, different island clusters including Pohnpei, Kosrae, Palau, Chuuk and Yap (Fig. 7.7) each had difficulty communicating with the other [18].

The trusteeship was treated as a war prize, paid for with the blood of the 32,000 Americans killed and maimed in the campaign across the western Pacific. The U.S. Navy administered the islands for a few years. Local teachers working in thatched huts established an education system. A network of trade stores was established and clothing was supplied, but consisted primarily of surplus military fatigues. A corps of medical officers and nurses were trained to introduce basic health care, at least on the main islands. American administration of the islands was generally opposed to forcing change on Micronesians [19]. However, that was not always the case, especially on the Marshall Islands where the U.S. decided to test nuclear weapons on the atolls of Bikini and Enewetak (formerly Eniwetok).

In February 1946, the United States government asked the 167 Micronesian inhabitants of Bikini Atoll to voluntarily and temporarily relocate so the United States government could begin testing atomic bombs for "the good of mankind and to end all world wars" [20].

The family heads were given to choose Rongerik Atoll (Fig. 7.7 inset) as their new home even though it was 200 km away, 1/6th the size of Bikini and had no fresh water. That is likely the reason why no one lived there, and just to make it more appealing, the traditional belief was that demon women haunted Rongerik. The U.S. military left the former residents of Bikini with a few weeks of food and water that soon proved to be inadequate [21] and began to busy themselves with the task of preparing Bikini for nuclear testing. The U.S. Navy brought in 95 ships of various sorts including retired aircraft carriers and battleships with tanks and other armored

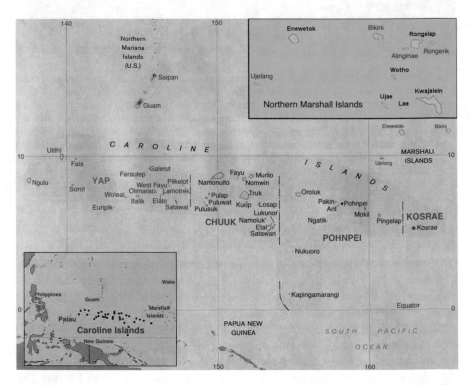

Fig. 7.7 Location of districts within Micronesia including the Mariana, Marshall and Caroline Islands. Palau, west of Yap is shown in the *lower* inset. *Upper* inset: expanded view of the northern Marshall Islands with atolls described in the text (Maps courtesy of Central Intelligence Agency and Wikipedia.org)

Fig. 7.8 Bikini Atoll lagoon test Baker just after detonation. The cloud ejected 1,500,000 cubic meters of seawater 1600 m into the atmosphere where it condensed and contaminated islands, groundwater, ships and sailors in 1946 (U.S. Government photo, Wikipedia Operation Crossroads Baker)

vehicles on deck to see how they might withstand the blasts. Live animals including more than 5000 goats, pigs, mice, rats and guinea pigs were positioned at varying distances from the ground zero for the same purpose [21]. Tens of thousands of sailors were stationed on ships 25 km or more from the first blast. However, there would be 23 such tests on various islands on the atoll between 1946 and 1958. Some were exploded in the air, but most of them were on barges, and one codenamed Baker, was detonated underwater. That test generated an enormous cloud of radioactive seawater that was thrust a mile high (1.6 km) in one second (Fig. 7.8). Once the cloud condensed it rained down across the atoll contaminating many of the islands as well as nearby ships and sailors. Some ships were so contaminated that they had to be sunk. There are no readily available accounts of the sailors who were exposed. The groundwater on larger islands became contaminated and proved to be a major stumbling block for resettlement by those islanders stranded on Rongerik [22].

Fission and Fusion Bombs

In the early years of the nuclear program atomic bombs were based on fission, the splitting of the nucleus of an element that generates two or more neutrons. These neutrons repeat the process, as they are captured by nearby atoms, thereby creating a chain reaction and an explosion. Only certain elements are readily fissionable. Uranium is one, but only when it is enriched with a high percentage of the specific isotope of uranium, U-235, as was the bomb dropped on Hiroshima. Plutonium as the isotope Pu-239 is also fissionable, and this was the type of bomb dropped on Nagasaki.

Fig. 7.9 The Bravo test crater is 2000 m in diameter and 76 m deep (Image courtesy of DigitalGlobe)

Fission bombs were constructed by separating a hollow U-235 or Pu-239 'bullet' in the tail from a cylinder of the same material at the front. A conventional explosive in the tail drove the two nuclear parts of the bomb together producing a critical mass and the chain reaction required for explosion. However, by 1952 the U.S. began testing bombs that drive hydrogen nuclei together as the first step. In that process called fusion, neutrons are captured and extreme heat is generated. The fusion fuel implodes as it ignites. Energy from the primary device then compresses a larger secondary section of the bomb containing both fission and fusion fuel. The combination produces many more neutrons and increases the explosive yield by at least tenfold.

The first practical version of the hydrogen bomb test was Castle Bravo in 1954. The yield was predicted to be 4–8 megatons (1 MT≈ 500,000 kg of TNT), but instead Bravo yielded 15 MT, or about 1000 times the power of all the bombs dropped on Japan during World War II. The blast vaporized two islands and part of a third (Fig. 7.9), and the superheated air may have contributed to a wind shift in the stratosphere that changed the direction of the fallout plume. Instead of going north as expected, the radioactive debris headed east, covering five other atolls, including two that were inhabited. One of those was Rongerik where Bikinians had been evacuated.

About 300 indigenous Marshall islanders inhabited Rongelap Atoll and they got some of the worst of the fallout (Figs. 7.7, 7.10). Some of its islands were covered with more than a cm of radioactive ash. Virtually all the inhabitants experienced severe radiation sickness including relentless vomiting and diarrhea, as well as burns, lesions, and other skin problems. After 3 days they were forced to abandon the islands and were taken for medical treatment. Three years passed before the United States government declared the area 'clean and safe' and allowed the islanders to return, though they were told to eat canned foods and avoid the northern islets

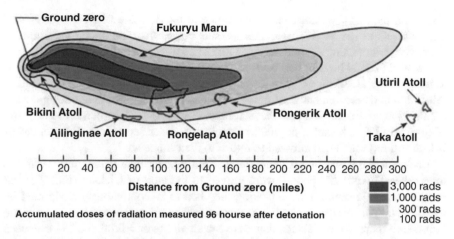

Fig. 7.10 Fallout pattern for the Castle Bravo test, 1954. Radiation Absorption Doses (RADS) delivered over a short period above 200 will cause serious illness; doses above 1000 (in *orange and red*) are almost always fatal (Image modified from Samuel Glasstone and Phillip J. Dolan (eds) The Effects of Nuclear Weapons, 3rd. edn. Washington, D.C.: DOD and DOE, 1977 and Wikipedia.org)

of the Rongelap where the heaviest fallout occurred. They were also told not to eat land crabs because those animals very efficiently recycle their calcium supply by eating their molted shells. Calcium is unfortunately mimicked by radioactive strontium-90 (half-life of 29 years[1]*) in the fallout. As a result land crabs were more radioactive in 1956 than they were earlier [23]. Despite precautions, evidence of continued contamination mounted as children developed thyroid tumors (from radioactive iodine), and many died of leukemia. According to one estimate, Castle Bravo rendered more than 10,000 km² uninhabitable, an area nearly the size of the state of Connecticut. Even on Bikini itself, radiation levels were dangerously high and American personnel were trapped inside their bunkers after the explosion [24]. In addition, the crew from a Japanese tuna boat, the Lucky Dragon, Fukuryu Maru (see Fig. 7.10), was covered with white radioactive ash causing all 23 crewmembers to become ill with radiation sickness.

The U.S. declared that Bikini was safe and encouraged return by providing free housing and food as inducements. About 100 islanders moved back in the mid-1970s and within a few years, low levels of plutonium were discovered in their urine. Internally, this radioactive isotope accumulates in vital organs and induces various cancers (see below). It is also pernicious externally due to its half-life of more than 24,000 years. Increases in the body burden of cesium-137 were also noted. That element that acts like potassium in the body and has a half-life of about 30 years. The Department of Energy suggested a complete feeding program to reduce dependence on local food, and hopefully, a reduction in exposure to radia-

---

[1] The half-life is the amount of time required for half of the amount of radioactive material originally present to decay. Half of Sr-90 for example will decay to non-radioactive zirconium in 28 years.

tion. That plan appears not to have worked and by 1978 the Bikinians developed higher levels of radioactive materials in their bodies than the maximum amounts permissible at that time, and all of them had to be re-evacuated. It was at that point, and after a Federal lawsuit had been filed, that the United States agreed to a radiological survey at Bikini [24]. In 1982, Congress provided the people of Bikini a relocation and resettlement trust fund. However, these funds were not designed for repatriation to Bikini. Instead these refugees were moved to certain islands on Majuro Atoll that began a population explosion described below. The people of Bikini still have not been allowed to return to their home atoll.

Enewetak Atoll and its residents fared slightly better than Bikini, at least in the long view. The island's people were moved to the atoll of Ujelang (Fig. 7.7) but because natural resources were lacking they had to be continuously resupplied by the U.S. Navy. Their living conditions were substandard, and became more so as the population increased on islands that were too small. As on Bikini, the 141 evacuees and their progeny became utterly dependent on the military for their existence. There were 43 nuclear tests on Enewetak, nearly twice as many as on Bikini. And in a similar fashion, some of the blasts, especially the hydrogen bomb tests, were more powerful than anticipated and contaminated the surrounding area. Some fizzled on the ground or on towers and failed to explode, but contaminated the site. Others were detonated and spewed radioactive fallout more broadly around the islands nearby [25]. The Enewetak people wanted to go home after the testing was completed, but the contamination levels were too high. During the late 1970s however, decontamination processes began. The military removed the topsoil on six different islands, mixed it with Portland cement and buried it in a blast crater created by 'Cactus' one of the last nuclear tests on the island of Runit. The crater was then covered by a dome made of 358 46 cm-thick concrete panels (Fig. 7.11). The cleanup costs were $239 million and the U.S. declared the southern and western islands of Enewetak safe for the return of the islanders, now 450 of them, in 1980. This was more than 30 years after they had been escorted from their homes. Runit Island would not be among those deemed inhabitable. Indeed it was quarantined forever because of the high concentrations of plutonium-239 buried within the dome and the residual plutonium still on the island [24].

United States military personnel were used to conduct the cleanup using bulldozers, trucks and other heavy equipment, but most were not issued protective gear. Some who were given the option were told that radiation levels were low and protective gear was unnecessary. Enewetak is tropical and hot. Most of the troops worked in shorts. Dust levels were high and covered everything during the cleanup. Workers were scanned daily for plutonium contamination and it was often at high levels. Showers removed the external material, but there was nothing that could be done for particles that were inhaled. Plutonium is most dangerous when it enters the lungs. The radiation can kill lung cells, and cause scarring, emphysema and cancer. From the lungs plutonium enters the blood stream and concentrates in bone, liver, spleen and elsewhere. Medical records for these soldiers are poor. Of the 4000 troops sent to clean Enewetak, many have died. Others are too sick to work and now (40 years later) suffer from bone disease including osteoporosis and tumors, as well as kidney, testicular and bladder cancers. As servicemen they cannot sue the mili-

Fig. 7.11 Runit Island at the northern side of Enewetak Atoll showing the seawater-filled blast crater of the Lacrosse nuclear test and the adjacent concrete-covered dome of the Cactus crater (Satellite image courtesy of DigitalGlobe, 2016. *Inset*: dome after construction in 1979. Image from U.S. Defense Special Weapons Agency and Wikipedia.org)

tary for medical care, and the Veterans Administration according to a report in the New York Times routinely turns down applications by individuals for service-related medical treatment [26].

As part of the cleanup, the existing coconut trees, whose coconut milk was contaminated with radioactive cesium, were cut down and replanted. The same was true of pandanus, breadfruit, and taro. It was found that the use of high potassium fertilizer competes with and reduces the uptake of cesium by plants and that was a method used during the remediation process [27]. In the meantime it was thought that U.S. food shipments would tide them over until the newly planted trees could begin to produce. As described in Sect. 7.7, that is not the way it worked out.

7.5   British and French Nuclear Tests and the Fallout Therefrom

The United Kingdom came to control the Line Islands south of Palmyra Atoll (Fig. 6.7) despite claims for several of them under the Guano Act by the United States. However, neither of these islands was occupied as required under the Act. Ultimately Malden, Christmas Island, and its neighbors were annexed by the United

Kingdom as part of the Ellice and Gilbert Islands Protectorate. That administration became a green light for the UK to test nine nuclear weapons in the atmosphere. Between 1957 and 1958, three were detonated at high altitude (helium balloons) over uninhabited Malden Island. The other six were air dropped over Christmas Island, named for Captain James Cook who arrived on Christmas day, 1777. However, the people who live in the region have difficulty pronouncing the letter s and thus Christmas became Kritmat, and that evolved phonetically into Kiritimati, an emergent atoll as described in Chap. 1. The main problem with Kiritimati was that at the time when it became a nuclear test site, some 300 people lived there. Much of the rest of the story is predictable. UK naval vessels evacuated the islanders during the tests. All of them and several thousand servicemen were positioned on boats or islands only a few kilometers from the test sites. That included a hydrogen bomb test that yielded three megatons, about 240 times as powerful as the one dropped on Nagasaki. Many were exposed to direct radiation from gamma rays. Fallout contributed to additional exposure as servicemen and the indigenous population were allowed to move around the island, drink and bathe in contaminated water, eat local fruits and vegetables, and breath radioactive dust. The military leadership largely ignored the dangers of radioactivity, forcing the Kiritimati people to file a lawsuit in the European Court of human Rights in 2006. Their claims alleged contamination and health problems that included thyroid tumors and high rates of leukemia, and malformations or stillbirths among their progeny. Nonetheless, no compensation was offered and no health studies have been undertaken. In addition, much of what was known about the aftereffects has been classified [28].

"If it's so safe, why not test it in Paris?" An Anti-nuclear Protest Sign Carried in Tahiti

In 1963 Britain, the United States and the Soviet Union signed a treaty that banned nuclear weapon tests in the atmosphere, in outer space, and underwater. Underground testing was still allowed, but France, although invited to sign, refused to do so. From 1960 to 1996, La République carried out more than 180 nuclear tests on the atolls of Mururoa and Fangataufa in the southeast Tuamotu Archipelago of French Polynesia (Fig. 5.4). More than 40 of these were conducted on the islands between 1966 and 1974. By 1968 only France and China were detonating nuclear weapons in the open air. The fallout from the blasts was widespread and led to multinational protests, especially among Pacific nations including Australia, New Zealand, Fiji, Samoa, Tonga and Tahiti. Some of the radioactive plumes were detected as far away as Peru. After 1975 France began underground nuclear tests in which devices were detonated in shafts that were drilled 750–1500 m deep into the volcanic rock under the atolls. Three of these were detonated on Fangataufa, but the other 120 were set off on Mururoa. Underground testing did little to quell the protests, especially when it was discovered that the atoll of Mururoa had developed a crack more than a 1600 m long and a meter wide that could release the radioactive material trapped in the blast shaft. In another case a nuclear device got stuck halfway down the shaft but was detonated nonetheless and caused a significant chunk of the atoll to break loose underwater. It also caused islands to sink several meters as the impact craters collapsed [29, 30].

With continued testing, protests occurred in a number of Pacific island countries and 13 of them joined forces with New Zealand for a Nuclear-Free Pacific in 1985 [31]. While the treaty prohibits the testing of nuclear explosives, that did not stop France from detonating another pair at Mururoa in 1995, just 1 year before the Comprehensive Test Ban Treaty was to be signed. Worldwide protest ensued. There were riots in Tahiti. French cafes in Australia had to change their names to disguise their identity, and wine shops reported that consumers were shunning French labels. Even in Paris, protest marches included nearly 3000 people [29, 31]. Quel horreur! The streets were said to have run red along with the faint odor of cheap Bordeaux. France's last nuclear test in the Pacific was a few months later in January 1996. However, the French government has steadfastly refused to compensate or even study nuclear workers and islanders who were exposed during any of their tests, although they built fallout shelters on nearby atolls and told the islanders there to take precautions [30]. Reports suggest leakage of radioactive material into Muroroa's lagoon, however the International Atomic Energy Agency and the government of France have concluded that no further environmental action, examination, or monitoring is required. Nonetheless, unauthorized visits to Muroroa are forbidden and the airstrip is guarded by a contingent of French Foreign Legionnaires [29].

7.6   The Postwar Period of Decolonization and Independence

Postwar independence movements were numerous on Pacific islands, but it was a slow process that began in the 1960s (Table 7.2). Full independence for the former Trust Territories of the United States was hampered by their strategic value on one hand, and their financial dependence on the other.

As a familiar theme among former colonies, the Marshallese accused the U.S. of broken promises including the right to return home and the decontamination of their home islands. They also complained that they were not being informed about the extent of their radiation exposure and petitioned the United Nations to end the testing. However the U.S. kept much of the information secret, assuring the world that there would be no lasting harm. After more than four decades under US administration as the UN Trust Territory of the Pacific Islands, the Marshall Islands attained independence in 1986, (Table 7.2) and became known as the Republic of the Marshall Islands (RMI). Under the terms of the Compact the United States and the RMI have full diplomatic relations. Marshallese citizens may work and study in the United States without a visa, and serve in the U.S. military. However, while the government is free to conduct its own foreign relations, it does so with the United States having full authority for its security and defense. The Department of Defense also has permission to use parts of the lagoon and several islands on Kwajalein Atoll for a missile test range (The Ronald Reagan Ballistic Missile Test Site) until 2066 with an option to continue until 2086. The Compact also provides for settlement of all claims arising from the U.S. nuclear tests conducted at Bikini and Enewetak Atolls from 1946 to 1958. For the loss of the homeland and the transformation of

Table 7.2 Postwar Island Independence [32]

Country	Former Administration	Year of Independence
Western Samoa	New Zealand	1962
Nauru	Australia	1968
Fiji	Britain	1970
Tonga	Britain	1970
The Ellice Islands become the nation of Tuvalu	Britain	1978
The Gilbert, Ellice and southern Line Islands become the nation of Kiribati	Australia, Britain	1979
Western New Guinea	Netherlands as Dutch New Guinea, then annexed as part of Indonesia	1963
Eastern (Papua) New Guinea incl. Bougainville Islands	Australia	1975
Solomon Islands	Britain	1978
New Hebrides Islands becomes Vanuatu	Britain and France[a]	1980
Marshall Islands	United States	1986
Yap FSM[b]	United States	1986
Chuuk, FSM[b]	United States	1986
Pohnpei, FSM[b]	United States	1986
Kosrae, FSM[b]	United States	1986
Palau	United States	1986

[a]An unusual form of government called a Condominium allowed both France and Britain to share colonial administration of Vanuatu from 1906 to 1980. There were separate British and French governments, including two immigration policies, two courts, two police forces, two sets of laws (British common law and French civil law), two health services, two education systems, two currencies, and two prison systems. Islanders were given the choice of which government they wanted to be part of until independence in 1980

[b]The Republics of Yap, Chuuk, Pohnpei and Kosrae are former Trust Territories in the Caroline Islands that are now independent nations, but are allied as members of The Federated States of Micronesia. All are part of the Compact of Free Association with the United States and each has agreements with the U.S. for aid and defense. The Marshall group and the Republic of Palau voted not to join the FSM, but are independently associated with the United States under their respective Compact Agreements

the Marshallese way of life caused by nuclear testing, the US government provided $150 million to an RMI compensation trust fund, under the condition that the Marshallese are prohibited from seeking future legal redress against the US. In addition the U.S. provides the Marshall Islands with approximately $70 million annually through 2023, including contributions to a jointly managed trust fund, as well as financial assistance in the form of other U.S. federal grants. Direct aid accounts for 60% of the Marshall Islands budget and the RMI is the largest employer; the U.S. Army is the second largest [33, 34]. The Marshall Islands, through no fault of their own, became a welfare state.

7.7 Dependency, Unemployment, Emigration, and an Homage to Spam

Because of contamination and disruption of native culture during the war, processed foods became a staple and took the place of traditional edibles. While by no means unique, the Marshall Islands provide the best illustration of that process. Some local foods such as crab, pandanus, local fish, and breadfruit are still consumed but when anything in a can is available, islanders are less likely to prepare edibles in the traditional way. Rice, introduced by the Japanese, has all but replaced taro as a starch source. Indeed, rice, tobacco and alcohol are among the three largest categories of imports to the islands [35]. More than 90% of food is imported, especially other carbohydrate-rich food such as pancakes, doughnuts and coffee bread that are consumed as regular meals. Between-meal snacks include uncooked ramen noodles and instant coffee with artificial cream and heaps of sugar. A recent study conducted in the capital of Majuro found that 53% of men and 70% of women are clinically obese, and blood glucose is abnormally high for 29% of men and 37% of women.[2] Marshall Islanders have some of the highest rates of Type 2 diabetes in the world and that disease is recognized as the top medical and public health problem, one that leads to secondary diseases from immune deficiencies, especially tuberculosis [36].

Tins of tuna, sardines and Spam have become the main source of protein for the Marshallese. Spam was first introduced to the islands during World War II. To Americans it was satisfactory at first, but when it began to turn up three times a day it became 'war food' and the butt of jokes. To the Marshallese, however, it became a delicacy and has become a comfort food as normal as breadfruit and fried reef fish. Vending machines in Majuro sell chips, candy bars and Spam [38]. This particular canned comestible was an invention of the Hormel Company. It was a means of dealing pork shoulder and other porcine parts that were discarded until someone at the company came up with the idea of using it for an inexpensive lunchmeat suitable for Depression-era pocketbooks. It debuted in 1937 as Hormel Spiced Ham, but that soon morphed into Spam. The military bought it not only because it was cheap, but also because it was easily transported and it was satisfactorily nutritious. According to the label the canned meat has only six ingredients: pork shoulder and ham, salt, sugar water, sugar, potato starch to keep the meat moist, and sodium nitrite as a preservative (Fig. 7.12).

In 1958 the Marshall Islands population was roughly 14,000, but increased U.S. subsidies in the mid-1970s led to the buildup of a commercial sector with modern department stores, restaurants, bars, movie theaters, and pool halls [35]. By 2014 the population had increased fivefold to nearly 70,000. Two thirds of Marshallese now live in Majuro and in the secondary urban center located on Kwajalein Atoll [39]. Such urban towns are already experiencing rising levels of unemployment,

---

[2] Obesity and diet-related diseases are not endemic to the Marshall Islands. In the Kingdom of Tonga, Samoa, Kiribati and the Federated States of Micronesia more than 50% of adults are obese [37]. The situation on Nauru is similar as described in Chap. 6.

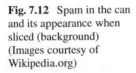

Fig. 7.12 Spam in the can and its appearance when sliced (background) (Images courtesy of Wikipedia.org)

particularly among the younger members of the workforce. In the Marshall Islands, young people are two to three times as likely to be unemployed as older residents; 20 percent unemployment is typical there, as well as in the Federated States of Micronesia generally. The high level of unemployment has prompted many in the Marshall Islands to leave. Although most of them in the United States reside in Hawaii, more than 6000 live in northwest Arkansas, with many working in the region's vast poultry processing industry. This migration to the U.S. is recent, increasing from 6700 in 2000 to 22,400 as of the 2010 census [40, 41].

References

1. Hiery HJ (1995) The Forgotten War: the German South Pacific and the influence of World War I. University of Hawaii Press, Honolulu
2. Ryden GH (1975) The foreign policy of the United States in relation to Samoa. Octagon Books, New York. (Originally published at Yale University Press, New Haven 1928)
3. CN MD, Gowdy JM (2000) Paradise for sale: a parable of nature. University of California Press, Berkeley
4. Fisher SR (2002) A history of the Pacific Islands. Palgrave Publishers, New York
5. O'Neill R (1993) Churchill, Japan, and British Security in the Pacific 1904–1942. In: Blake RB, Louis WR (eds) Churchill. Clarendon Press, Oxford
6. Myers RH, Peattie MR (1984) The Japanese Colonial Empire, 1845–1945. Princeton University Press, Princeton
7. Rottman GL (2003) Japanese Pacific Island Defenses 1941–1945. Osprey Publishing, Elms Court
8. Midway Atoll National Wildlife Refuge https://www.fws.gov/refuges/profiles/History.cfm?ID=12520
9. Bennett JA (2009) Natives and Exotics World War II and environment in the Southern Pacific. University of Hawaii Press, Honolulu
10. Fisher HI (1949) Populations of birds on Midway and the man-made factors affecting them. Pac Sci 3:103–110

11. Pacific Islands Benthic Habitat Mapping Center http://www.soest.hawaii.edu/pibhmc/pib-hmc_pria.htm#johnston
12. Seelye K (2003.) Radioactive dump on Pacific wildlife refuge raises liability concerns New York Times 27 January, 2003. http://www.nytimes.com/2003/01/27/us/radioactive-dump-on-pacific-wildlife-refuge-raises-liability-concerns.html
13. Dawson EY (1959) Changes in Palmyra Atoll and its vegetation through the activities of man, 1913–1958. Pac Nat 1:1–51
14. Gardner PA, Garton DW, Collen JD (2011) Near-surface mixing and pronounced deep-water stratification in a compartmentalized, human-disturbed atoll lagoon system. Coral Reefs 30:271–282
15. Maragos J, Friedlander AM et al (2008) US coral reefs in the Line and Phoenix Islands, status, threats and significance. In: Riegl BM, Dodge RE (eds) Coral Reefs of the USA. Springer Science + Business Media BV, Dordrecht
16. Collen JD, Garton DW, Gardner JPA (2009) Shoreline changes and sediment redistribution at Palmyra Atoll (equatorial Pacific Ocean): 1874–present. J Coast Res 253:711–722
17. JPA G, Bartz RJ et al (2014) Conservation management options and actions: putative decline of coral cover at Palmyra Atoll, northern Line Islands, as a case study. Mar Pol Bull 84:182–190
18. Kluge PF (1991) The edge of paradise: America in Micronesia. Random House, New York
19. Hezel FX (2001) The new shape of old island cultures. University of Hawai'i Press, Honolulu
20. Niederthal J (2001) For the good of mankind: a history of people of Bikini and their Islands. Micronitor/Bravo Publishers, Majuro
21. Jacobs RA (2010) The dragon's tail: Americans face the atomic age. University of Massachusetts Press, Boston
22. Maragos JE (2011) Bikini atoll, Marshall Islands. In: Hopley D (ed) Encyclopedia of modern coral reefs, structure, form and process. Springer Science + Business Media BV, Dordrecht, pp 123–136
23. McCool WC (1957) Return of the Rongelapese to their home island. Report from the Atomic Energy Commission. https://web.archive.org/web/20070925185914/http://worf.eh.doe.gov/ihp/chron/A43.PDF
24. Johnson G (1980) Paradise lost. Bull Atom Sci 36:24–29
25. Maragos JE (2011) Enewetak atoll, Marshall Islands. In: Hopley D (ed) Encyclopedia of modern coral reefs, structure, form and process. Springer Science and Business Media BV, Dordrecht, pp 380–391
26. Philipps D (2017.) Troops who cleaned up radioactive islands can't get medical care. https://www.nytimes.com/2017/01/28/us/troops-radioactive-islands-medical-care.html?_r=0
27. Danesi PR (2009) Remediation of sites contaminated by nuclear weapon tests. In: Voigt G, Fesenko S (eds) Remediation of contaminated environments. Elsevier, Amsterdam, pp 223–262
28. https://www.ctbto.org/specials/testing-times/8-november-1957-grapple-x
29. Stanley D (2000) South Pacific handbook, 7th edn. Avalon Publishing, Emeryville
30. Brij VL, Fortune K (2000) The Pacific Islands, an Encyclopedia. University of Hawaii Press, Honolulu
31. Siskin KM (1996) Does international law reflect international opinion? French nuclear testing in the twentieth century. Georgia J Intl Comp Law 26:187–214
32. Thompson RC (2001) The Pacific Basin since 1945. Routledge, London
33. U.S. Department of State Fact Sheet 2016, U.S. relations with Marshall Island [sic]. https://www.state.gov/r/pa/ei/bgn/26551.htm
34. Wong A, Blair C. (2011.) Compact of Free Association Honolulu Civil Beat (May 26, 2011). http://www.civilbeat.com/topics/compact-of-free-association
35. Hanlon D (1998) Remaking Micronesia. University of Hawaii Press, Honolulu
36. Ahlgren I, Yamada S, Wong A (2014) Rising oceans, change, food aid and human rights in the Marshall Islands. Health and Human Rights Journal (online vol. 16). https://www.hhrjournal.org/2014/07/rising-oceans-climate-change-food-aid-and-human-rights-in-the-marshall-islands/

37. Non-Communicable Disease Roadmap 2014. World Bank Joint Forum on Economic and Pacific Health. http://documents.worldbank.org/curated/en/534551468332387599/pdf/89305 0WP0P13040PUBLIC00NCD0Roadmap.pdf
38. Rudiak-Gould P (2009) Surviving paradise, one year on a disappearing island. Union Square Press, New York
39. CIA (2014) The world factbook – Marshall Islands. Available at https://www.cia.gov/library/publications/the-world-factbook/geos/rm.html
40. http://www.migrationpolicy.org/article/marshall-islanders-migration-patterns-and-health-care-challenge
41. https://siteresources.worldbank.org/INTPACIFICISLANDS/Resources/Chapter%2B1.pdf

Chapter 8
Compound Issues of Global Warming on the High and Low Islands of the Tropical Pacific

Abstract The evidence for climate change and global warming, and their relationship to greenhouse gases is briefly reviewed. The terrestrial realm of Pacific high islands is described with particular emphasis on New Guinea along with the combined impacts of natural resource extraction and climate change. The high islands of Hawaii serve as additional examples. Low islands are often limited by water for drinking and agriculture. This resource is more severely strained on highly populated islands where there are often insufficient amounts of it, as well as increased levels of contamination. Rising sea levels exacerbate the problem of limited water resources, especially on low islands, as does increasing frequencies of El Niño and La Niña cycles. The Solomon Islands in the western Pacific are identified as a rising sea-level hotspot due to their proximity to the Pacific Warm Pool. The effect of warming water on tuna resources and their migration patterns are described. Coral reef bleaching and its frequency as a global phenomenon are presented. Coral reefs everywhere have experienced deadly levels of warm and acidified water that threaten one of the most spectacular and diverse ecosystems on Earth.

There are some ten million people in 22 countries in the tropical islands of the Pacific Ocean. Melanesian islands are the most populous with nearly nine million people and correspondingly have the largest land area. Polynesia has a population of about 656,000 and Micronesia has an estimated 507,000 people as of 2016 [1]. The high islands and low islands of each region represent distinctive habitats and challenges as population pressures strain resources and cause environmental degradation. This set of problems will be explored comparatively in this chapter, but there are compounding influences due to climate change that will be examined first.

8.1 Climate Change and the Atmosphere

Some background, principles and definitions are necessary to understand the nature of climate change. Principle number one: Climate is not the same as weather. Weather patterns include shifts in temperature, rainfall and wind that occur in the short term. When such conditions have a characteristic pattern that occurs repeatedly over decades or more, that is the climate. A Mediterranean climate for example, is characterized by warm, dry summers and cool, wet winters. This category is typical of the western portions of continents between about 30° and 45° latitude north and south of the Equator. The widely used Köppen classification scheme divides climate into five major groups: tropical, dry, temperate, continental and polar. Each of the five is further divided into multiple subcategories. The Mediterranean, for example, is one of nine subcategories of temperate climates.

Principle number two: Climate change and global warming are not the same. Global warming refers to the recent and ongoing rise in global average temperature near Earth's surface. Increasing concentrations of greenhouse gases in the atmosphere, which trap heat in the atmosphere much like the glass in a greenhouse, are the primary contributors to this warming trend. More critically, the volume of these gases, CO_2 in particular, has become especially notable within the last 50 years. During this short interval, humans have increased their numbers by 68% and have been burning an increasingly greater quantity of fossil fuels including oil, gas, coal and peat, as well as solid waste and wood. In addition to combustion, certain types of manufacturing including cement production and the raising of livestock are also significant sources of greenhouse gas emission. Because trees take up carbon dioxide and accumulate it as wood and other storage forms of carbon, forest conservation could be an antidote for CO_2 accumulation. However, during the first 12 years of this century Earth lost a net of 1.5 million km^2 (580,000 square miles) of forest. That is more than the land area of Alaska deforested by fire, logging, disease and insect pests [2]. Loss by fire is due to longer dry seasons and changes in long-term rainfall patterns; insect pests such as bark beetles destroy increasingly large tracts of pine forests that at one time were too cold for the beetles to produce viable eggs. There are fewer such refuges now.

The atmospheric greenhouse is natural and necessary to support life on Earth. Water vapor in the atmosphere (water in the gaseous phase) is by far the largest contributor to the greenhouse effect. However, its lifetime is about 9 days after which it condenses and falls back to Earth as precipitation [3]. Some gasses are better at producing a greenhouse than others, either because of its amount, atmospheric lifetime, or relative ability to retain heat (Fig. 8.1).

Carbon dioxide (CO_2) is the most abundant of these gases and it remains in the atmosphere for hundreds of years. Furthermore, the amount of it has increased by nearly 20% since 1980. Because of its rapidly increasing abundance and long residence time, carbon dioxide reductions, if they occur, will only slowly diminish and gradually reduce its heat-trapping effect. The Arctic is an enormous source of carbon, the release of which is promoted by a warming Earth. Its permafrost is, as the name suggests, permanently frozen ground that is rich in soil and plant material that has not undergone bacterial decomposition due to its frozen state. However, a vast expanse of permafrost in Siberia and Alaska has started to thaw for the first time since it formed more than 100,000 years ago. In the rapidly warming Arctic, which is heating twice as fast as the globe as a whole, the upper layers of this frozen organic material is beginning to decompose [4]. The problem is that permafrost contains almost twice as much carbon as is now present in the atmosphere.

Human Influence on the Greenhouse Effect

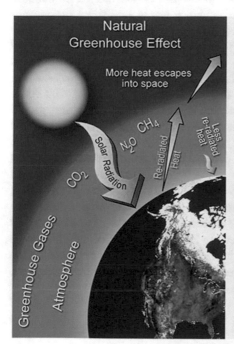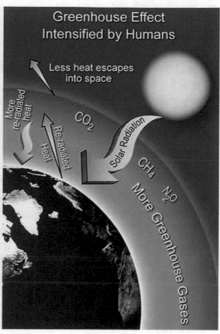

Fig. 8.1 Illustration of the greenhouse effect due to increases in atmospheric greenhouse gasses including carbon dioxide (CO_2), methane (CH_4) and nitrous oxide (N_2O) (Courtesy of William Elder, National Park Service National Climate Assessment, 2014 U.S. Global Change.gov)

By contrast, methane (CH_4) remains in the atmosphere for much less time (six to seven decades) [5], but it can trap 30 times the heat ton-for-ton compared with CO_2. The Arctic is rich in frozen methane, called methane hydrate that is found either in the permafrost or in the sediments of shallow seas. There is additional methane that is trapped by association with carbon. Similar to permafrost, this organic form of methane will be released when warming allows bacterial decomposition to take place [6]. Estimates suggest that methane release from the Arctic will double the amount currently present in the atmosphere. A third example of greenhouse gas emission is nitrous oxide N_2O, a natural component of the atmosphere derived from microbial activity in soils and in the ocean. Human production of nitrous oxide is primarily due to combustion of fossil fuels, fecal production by livestock, industrial production of nitric acid, and application of fertilizers to agricultural crops. This gas makes up a very small amount of the greenhouse volume but it remains in the atmosphere for 114 years and is about 300 times more effective at trapping heat compared with CO_2. During its atmospheric residence, N_2O is broken down by ultraviolet light, and that would be fine if those nitrogen-containing breakdown products did not combine with ozone in the stratosphere [7]. That combination causes ozone depletion, which in turn leads to high ultraviolet light exposure and an increase in the incidence of various skin cancers.

Principle number three: There is nothing natural about the current warming trend. Earth has experienced warm periods in the past before there were humans. We know about past climates because of evidence left in tree rings, ocean sediments, coral growth, layers of sedimentary rocks, and layers of ice in glaciers. Ice layers represent annual snowfalls that trap air bubbles from the atmosphere as they are frozen. Thus each layer of ice taken from a core contains a record of atmospheric gases at the time of its formation. In Antarctica some cores are more than 4600 m long and give us an atmospheric record that goes back more than 800,000 years. That includes a record of greenhouse gasses and global temperatures. Cold, glacial periods that affect polar to mid-latitudes, persist for about 100,000 years and these intervals are punctuated by briefer, warmer periods called interglacials. The current interglacial is the Holocene and it began about 11,700 years ago. But the last time it was as warm as it is now was during the Eemian stage, the last interglacial period of the Pleistocene that began approximately 130,000 years ago and ended about 115,000 years ago. The warmest part of the Eemian developed over thousands of years and peaked between 127 and 122 thousand years ago [8], and other similar warm and cold intervals that characterized the 2.6 million years of the Pleistocene epoch exhibited a very similar pattern. Interestingly, the global temperatures we see now are less than 1 °C from those of the Eemian and other interglacial warm periods, as shown in Fig. 8.2 [9].

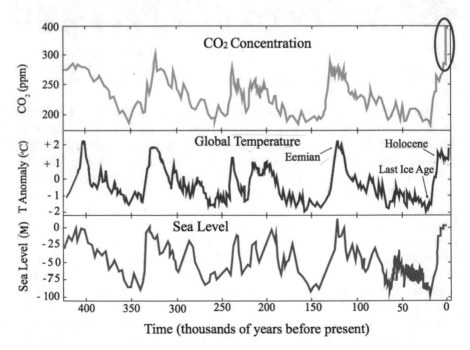

Fig. 8.2 Global average temperature (°C), carbon dioxide concentration, and sea-level rise and fall in meters during the last 450,000 years, including the Eemian and the present (Graph courtesy of John Englander, adapted from Hanson and Sato [9])

This ongoing cycle of glacial and interglacial periods closely matches variations in Earth's orbit around the sun as well as its axial tilt, and changes in its rotational axis. It has been suggested by some that this is just another such cycle, except that these changes occur at 40,000–100,000 year intervals. During the last global warm period, carbon dioxide peaked during the warm climate of the Eemian, as it did during every cycle of increased temperatures (Fig. 8.2). However, during the entire span of the Pleistocene, CO_2 was never as high as it is now. To get to current levels, one would have to go back 3.3 to 3.0 million years to the mid-Pliocene warm period when there were no Arctic or West Antarctic ice caps, global temperatures were at least 2°C warmer than now and sea level was approximately 23 m higher [10, 11].

Moderate climate oscillations such as the Little Ice Age and the Medieval Warm Period occurred during the last millennium. However, while the forces that gave rise to these changes are complex, the evidence suggesting that these events extended beyond the northern hemisphere is not widely accepted (but see Ref. [12] for a contrarian view). Indeed, the temperature increase associated with the Medieval Warm Period that occurred approximately between 900–1300 AD was regionally concentrated, and while some portions of the northern hemisphere were as warm then as they are now, other areas were relatively cool. The tropical Pacific in particular was exhibiting strong La Niña cooling during this period [13].

The recent warming of the Earth as a whole is not only supported by temperature measurements. Climate changes include major shifts in precipitation, wind patterns, and sea levels among other effects that occur over several decades or longer. During the last 30 years the Arctic ice cap has retreated by more than 17% per decade as compared with losses of about 4% per decade through the 1980s [14]. Thus, ice cap losses are accelerating and during the last 20 years the amount of snow and ice cover across the Arctic has been cut in half. Summer sea ice in the Arctic Ocean disappeared along the Russian coast for the first time in 2012 and all of it may be gone by 2030 [15]. Losses are also occurring in the ice sheets on the Antarctic Peninsula. On the western ice shelf the losses are obvious with chunks the size of small states breaking off and floating away during the last decade. Even though snowfall on the western Antarctic ice shelf is continuing, the losses in that region are expected to outpace the gains within a few decades [16]. Losses from the eastern Antarctic ice shelf, by contrast, are fairly subtle because satellite images fail to show the thinning process that occurs by warm ocean water eroding the ice cap from below [17]. There is also an unprecedented and unmistakable decline of glaciers throughout the world from New Zealand and the Andes to Alaska, Greenland, Europe and the Himalayas [18].

More winter and spring rain is projected for the northern United States leading to frequent flooding; there will be a decrease in rainfall in the southwest over this century, resulting in extreme heat and drought, insect outbreaks and reduced agricultural output. The frequency and intensity of storms has increased since the 1980s and sea level is expected to rise more than a meter by the end of the century, although the exact number depends on what is done to contain greenhouse gasses [19]. If it is 'business as usual' the number may be closer to 2 m, mostly from shrinking of polar glaciers and the feedback mechanisms that accelerate the process, such as reflected light. As ice loss continues, less light is reflected, more heat is absorbed at the Earth's surface, and the melting rate increases.

The National Flood Insurance Program provides updated maps that determine risk of flooding for the United States. Created by the Federal Emergency Management Agency (FEMA), these maps are revised because flood risks change over time. In some cases those changes may be local due to land use and development, or because of changing regional weather patterns. By contrast, sea-level rise is none of the above. Today coastal communities are seeing more frequent flooding during high tides, and this pattern will continue over the next 15–30 years as the rate of sea-level rise accelerates due to melting land ice, and warming which causes thermal expansion of water. Local topography and oceanographic features ensure that these effects will not be the same everywhere. What is generally referred to as 'sunny-day flooding' or sometimes 'nuisance flooding' (as if it were nothing more than an inconvenience) is on the increase, especially on the east coast of the United States from Maine to Texas [19]. In Charleston, South Carolina for example, such floods occurred 2–3 days per year in the 1970s [20]. Now, however, sunny-day floods occur ten or more days per year. Similar patterns have been noted from Miami Beach to Boston and they are especially noticeable during so-called king tides that occur once or twice a year when the Earth, Moon and Sun are closest to one another in their orbits. King tides simply exacerbate normal high tides that occur on a lunar cycle, and any concurrent storm surge worsens the effect. The city of Miami Beach, which is only about a meter above sea level, has found itself underwater more frequently lately than they would like, especially in the last decade in conjunction with sunny days during king tides [21]. The Sun and Fun Capital of the World is spending as much as $500 million to raise roads and seawalls across the city. They are also installing 80 massive water pumps to combat the flooding. Inland cities near large rivers also experience more flooding, especially in the Midwest and Northeast. Hotter and drier weather as well as earlier snow melt mean that wildfires in the West start earlier in the spring, last later into the fall, and burn more acreage [22]. It may be worth noting that 16 of the 17 warmest years on record globally have occurred since 2001.

8.2   High Island Environments

The Terrestrial Realm of New Guinea and Other Western Pacific Islands

New Guinea is the largest island in Melanesia and in the tropical Pacific. It is divided in half politically, with the western side known as West Papua, belonging to Indonesia. The eastern half is the independent nation of Papua New Guinea, which includes the high islands of the Admiralty and Bismarck Archipelagos as well as a series of low islands and atolls, especially to the north and the east of the main

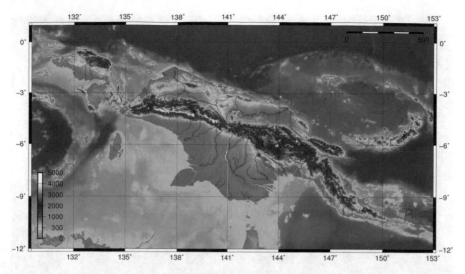

Fig. 8.3 New Guinea topographic map showing mountain regions in red and white; lowland areas comprising rainforest, savannahs and river deltas are depicted in green. The island is divided into an eastern portion comprising the independent state of New Guinea (PNG), which includes the Bismarck Archipelago shown at upper right. The remaining half is West Papua, part of Indonesia. In reference to text section 8.6: The Gulf of Papua is a well-developed mangrove area between 144° and 147° longitude at the southeast end of the island, as is Bintuni Bay in the Bird's Head Peninsula in the northwest (upper *left*, −2°, 133°). Reefs are well developed along the southeast of Papua New Guinea, an area known as the Bird's Tail Peninsula, as well as along the Bird's Head (Map courtesy of Wikimedia.org)

island [23]. The island is culturally and physically diverse with high mountains of the interior reaching close to 4600 m (Fig. 8.3). Snow has been known to fall in the higher elevations even though they are located near the equator. At the beginning of the twentieth century, at least five glaciated regions remained on the highest mountains, although most of them are now gone.

Natural Resources and their Extraction

About 65% of the island is covered by forest, much of it pristine, including rainforest near sea level and montane forests at higher altitudes. These native woodlands are also home to more than 780 species of birds, 10% of which are found nowhere else on Earth [24]. There are also kangaroos that live in trees, fruit bats with 1.6 m wingspans, and Queen Alexandra's birdwing, a butterfly 25 cm across. Between 1972 and 2002 a net 15% of Papua New Guinea's tropical forests were cleared for numerous purposes described below. However logging itself has degraded a substantial amount of forestland. Current practices target large, old-growth timber, but clear cutting is also employed and regulations prohibiting it are difficult to enforce.

Fig. 8.4 Palm oil plantation after replacement of native forest as seen from an altitude of 2600 m, Eastern Highlands Province, Papua New Guinea (Image courtesy of DigitalGlobe)

The result is deforestation, which includes not only the removal of vegetation and topsoil in the logged area, but also affects former forestlands that were turned into access roads, as well as the areas crushed by the dumping of logs where they were loaded for pickup [25]. The scrubby regrowth that follows bears no resemblance to the original habitat, and logged forests thereby support little of the diversity they once had. Forests are also prone to wildfires, especially during droughts (see below), or they may be purposefully set afire to create farmland. Large-scale forest clearance for oil palm production (Fig. 8.4) is becoming more common along with the accompanying pesticide use, and pollution in various forms. Until recently, such plantations have been only a minor contributor to forest loss, with subsistence agriculture and logging as the primary forces. However, there are now nearly a million hectares of proposed oil palm developments in Papua New Guinea as traditional land ownership is converted to long-term corporate leases [26].

Because the islands of the tropical SW Pacific lie along the subduction zones between the Pacific and Australian plates, they are tectonically and volcanically active. New Guinea for example is prone to earthquakes, 20 having recorded magnitudes of 7.0 or more in the last century. These have caused landslides, successive aftershocks, and high waves, including some that have generated severe tsunamis. There are also numerous volcanoes, most of which are inactive, but 16 have erupted in the last century, five in the last 15 years on the island of New Britain in the

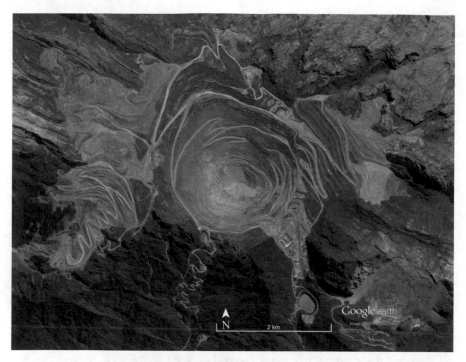

Fig. 8.5 The Grasberg gold and copper mine in the mountains of West Papua, altitude 4300 m. The open pit at center is 1.6 km in diameter. (Image courtesy of DigitalGlobe)

Bismarck Archipelago alone. The large physical area and its tectonic history additionally provide numerous natural resources and opportunities for their exploitation. Oil and gas were discovered in the 1980s and still make up the island's largest export item. Extensive drilling is occurring and pipelines have been constructed by multinational corporations in the highlands, as well as offshore in the Gulf of Papua (Fig. 8.3) where low labor cost and proximity to the Asian markets are cited as major incentives for continued development. Exxon Mobil for example intends to invest $19 billion in a liquid natural gas project that began in 2014 in the Southern Highlands Province. While there are multiple fields and a project lifetime of 30 years, there are also issues involving revenue sharing with clans in the area, and the translation of these investments into the establishment of infrastructure and education. These problems are not endemic to PNG.

Gold has long been sought in the region, beginning with the Portuguese discovery of the Solomon Islands, named for the Biblical king and the treasures thought to be there. Indeed gold was mined in Guadalcanal during the 1990s although Western concepts of land ownership did not sit well with Melanesian society. In addition, ethnic tensions having nothing to do with mining caused the mines to close, reopen and close again [27].

By contrast, significant amounts of gold, copper, silver, and other minerals have been extracted in extensive mining operations in New Guinea since the 1970s.

Three of these are top producers even by global standards, and contribute significantly to the economies of West Papua and Papua New Guinea, although at a price. Several of these including the Grasberg Mine (Fig. 8.5) are open pit operations occupying high-relief land that is subject to substantial rainfall, landslides, and river contamination. The OK Tedi mine near the border with West Papua, for example, discharged 40 million metric tons of mine tailings (waste rock and fine particles remaining after ore extraction) and other wastes into local rivers each year of its operations [28]. Fish populations became severely depleted, previously arable land became buried, and most of the forest birds along the river corridors migrated [29]. Environmental lawsuits were filed and arguments concerning corporate responsibility continued for years until the PNG government granted prosecutorial immunity to the mine owners and in 2013 took over the mine.

8.3   Effects of Climate Change on High Islands

Papua New Guinea

There are two types of oceanic rainfall that can be distinguished on islands. *Convective rain* occurs everywhere, including over the open ocean and on islands of all kinds when air rises above the ocean surface and condenses to form clouds and precipitation. In contrast high islands experience *orographic rainfall* when warm humid air is forced to rise up on mountain slopes. There the cool air it encounters causes the water vapor to condense into clouds and rain. As a high tropical island, Papua New Guinea experiences considerable orographic rainfall, but it is discontinuous due to seasonal shifts in the trade winds and positions of the intertropical convergence zones (Chap. 1). Shifts in rainfall may also occur periodically due to the occurrence of El Niño and La Niña oscillations that have different effects depending on geographic position. El Niño in Papua New Guinea promotes a shift in the position of the Pacific Warm Pool (Fig. 1.1) toward the central Pacific. This brings higher-than-normal rainfall to Kiribati and other Pacific islands farther east that are relatively dry. That warm water shift induces drought, crop failures, and fires in PNG. Conversely, La Niña is sometimes referred to as El Niño's soggy sister, due to the higher levels of rainfall than is normal for PNG. Higher rainfall from La Niña conditions in New Guinea means that islands of the central Pacific and points farther east will be subjected to drought. The effects of climate change layered atop these cycles is expected to intensify their effects, but there is a very high degree of confidence that the direction of long-term change for Papua New Guinea will result in an increase in mean temperatures, and in the frequency and intensity of extreme rainfall [30]. These events are expected to affect river, stream, and coastal flooding, as well as road and bridge washouts and additional effects described below.

Palm oil production is expected to benefit from higher amounts of rainfall. Oil palms also do well under increased conditions of carbon dioxide and they also have a broad temperature optimum [31]. Likewise, cocoa and coffee production may expand, but diseases that affect these crops are also expected with higher humidity. Increases in atmospheric carbon dioxide will benefit maize, sorghum and sugarcane production because these are warm-weather crops that are efficient at gathering carbon dioxide and converting it into a four-carbon form during photosynthesis. These so-called C4 plants also excel at utilizing nitrogen from the atmosphere and in the soil, and because they use water efficiently, they do well in warm and sunny environments. By contrast, bamboo, conifer and beech forests, along with highland ferns and orchids grow in PNG's cool and wet mountain regions. These are primarily C3 plants that do well under native conditions, but will likely diminish with warmer weather and higher levels of carbon dioxide. Likewise, rubber farming is not compatible with high rainfall that washes latex from collection containers. In addition, the transmission of disease to humans is expected to increase as weather warms in the highlands, favoring the migration of *Anopheles* and other mosquitoes as well as their increased activity in the transmission of malaria and dengue fever [32]. Extreme rainfall and flooding can also promote cholera and dysentery, diseases that are known to increase after flooding events contaminate food and drinking water with fecal waste, as will likely occur in the coastal and estuarine regions of PNG.

The High Islands of Hawaii

The southeastern Hawaiian Islands are the largest of the archipelago, and six of them reach altitudes of over 1000 m. These are Hawaii, Maui, Oahu, Kauai, Molokai and Lanai from largest to smallest. Ocean temperatures differ more on a seasonal basis here due to their position 19–22° north of the equator compared with New Guinea's equatorial location. Nonetheless, Hawaiian waters have been exhibiting a decade-to-decade warming trend since the 1950s, and an increased frequency and severity of coral bleaching, as described below. El Niño warming events also affect the islands, and the one in 2015 caused some of the worst coral bleaching ever recorded in the State. A warming climate also affects freshwater resources but in a manner that is dependent on island size and topography. However, on most islands, increased temperatures coupled with decreased rainfall will cause draughts that are expected to reduce the amount of freshwater available for drinking and crop irrigation [33].

Surface air temperature is expected to continue to rise over the entire region, especially at high elevations. Existing climate zones on high islands are generally projected to shift upslope and invasive plant species with a tolerance to higher CO_2 levels are expected to invade higher elevation habitats as climates warm. Likewise, Hawaiian high-elevation alpine ecosystems on Hawaii and Maui islands are warming in ways that affect avian populations. The lowlands on the islands were once home to 100 species, including several endemics. Hunting, introduced predators, and habitat destruction have reduced that number to 48, eleven of which have not been seen in decades and are probably extinct [34, 35]. Hawaii's spectacular and unique native songbirds known commonly as honeycreepers, are now largely limited to high-elevation forests. However, their formerly safe montane habitat is increasingly vulnerable as rising temperatures allow mosquitoes carrying diseases such as avian malaria to thrive at higher elevations [34, 36]. Ironically, Hawaii was mosquito free until American and European whaling and trading vessels introduced them in the early nineteenth century.

8.4   Climate Change on Low Islands

Most of the tropical Pacific islands in Micronesia and Polynesia are small, although some develop elevation with the characteristics of high volcanic islands. For example Yap in the Caroline group is 120 km^2 and is 178 m high, similar to nearby Kosrae, and Rarotonga in the Cook Islands (Fig. 7.7). The island of Savai'i in Samoa is among the largest of the region at 1718 km^2, and boasts one of the highest peaks in Polynesia at 1858 m [37]. Still others, such as Bora Bora in French Polynesia, are partly volcanic with a 727 m-high central peak, and partly limestone developed by reef growth and sediment accumulation (Fig. 1.11b).

Low Islands are typically those formed by the latter processes of organic growth and accumulation, and although there are some that have been uplifted 60–90 m above sea level (e.g., Makatea, Nauru, Niue and Banaba) (Fig. 1.0), most in that category are islands and atolls that are 3–4 m above sea level (Chap. 1). In contrast to high islands where surface water in the form of streams, springs, lakes, and swamps may be found, fresh water does not always occur on low limestone islands, and they may remain uninhabited as a result. In some areas, rainwater is collected on rooftops or from runways as it is on the atolls of Tuvalu and those of the northern Cook Islands, where that may be the only freshwater available. In other areas, groundwater may occur as a limited natural reservoir (an aquifer). In some cases the presence of an aquifer depends on an impermeable rock layer that holds rainwater above the water table. In other cases where there is no impermeable layer, rainwater percolates through porous rock and floats as a so-called lens over a layer of denser brackish or salt water (Fig. 8.6).

The volume of freshwater lenses is roughly proportional to the width and surface area of an island and is also influenced by the amount of rainfall and limiting factors such as salt mixing due to storms or even from high tides. On some islands the lens may be as thick as 20 m, whereas on many low-lying atolls, they may only extend to 10–20 cm. Lenses on the larger islands of Kiribati are the primary source of freshwater, although supplemental rainwater supplies are typically used when it is available. However, water is almost always limited on some of the smaller islands in Kiribati and elsewhere; many of the Marshall Islands, for example, require water supplies to be brought in regularly by boat. During severe droughts or natural disasters, residents may have to rely on coconuts for their water supply [39].

Population pressure is increasing the demand for water and puts a strain on the fragile lens supply. That pressure is expected to increase over the next few decades. At the same time, urbanization is increasing on certain islands, which diminishes water supplies by pollution and contamination of groundwater. This is well illustrated

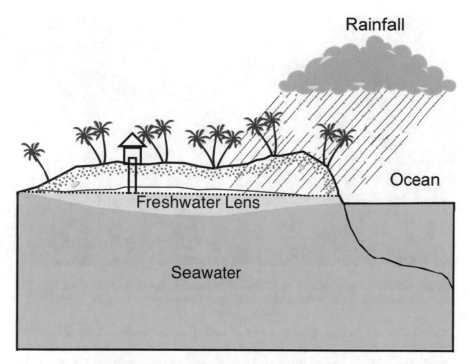

Fig. 8.6 Rainfall supplies and recharges an island freshwater lens. After percolating through sandy sediment or porous rock, freshwater lies atop the denser seawater (Modified from [38] and U.S. Global Change Research Program)

by the Republic of Kiribati, which includes the Gilbert Islands, especially Tarawa, an atoll where nearly half of the nation's people are crowded onto two islets. One of these is Betio where in 1940 the population was roughly 400; it is now more than 50,000 and there are overwhelming concerns about the supply of freshwater (Fig. 8.7). Water use for traditional crops such as coconuts, pandanus and taro often competes with drinking water supplies. In addition, poverty, overcrowding, latrines and septic tanks, pesticide or fertilizer use, and domestic animals, pigs in particular, all contribute to groundwater pollution and waterborne diseases [39]. Thus for Tarawa and other population centers like it the problem is twofold, not enough water and not enough safe water.

Rising Sea Level

The effects of global sea-level rise on atoll environments have been in the news for some time, with headlines suggesting that low islands are disappearing. However a 3 m-high ridge built by waves on the windward side fronts most atolls across the Pacific. This feature indicates a net gain rather than long-term erosion [40].

Fig. 8.7 Satellite view of Betio Island, 2016. High elevation here is 3 m above sea level. The population density is on par with that of Hong Kong or Tokyo (Courtesy of DigitalGlobe)

Indeed historical photographic analyses of atoll islands including those of Funafuti, the capital of Tuvalu (formerly the Ellice Islands) and Tarawa have suggested that islands are dynamic and resilient, and may change shape due according to oceanographic conditions. Some islands appear to have lost area here or gained area there, but they are generally persistent features. Indeed, Betio Island on Tarawa appears to have gained 30% of its island area since World War II [41]. These results have suggested a more optimistic prognosis for atolls and their ability to cope with rising sea levels [42]. However, in terms of Betio and other islands on southern Tarawa, much of their increase in size appears to be driven by dredge and fill projects that created land, or by seawalls and similar artificial structures that changed the natural flow pattern of shoreline sediment [43].

Seawalls are most prominent in urban and surrounding areas where population pressures are greatest [44]. Additional human-induced sources of erosion here include nearshore reef destruction from terrestrial pollution, and removal of an estimated 70% of mangrove forest compared with the 1940 baseline [45]. Such problems are discussed further in section 8.6. Sea-level rise is critical for low islands including hundreds of inhabited atolls that at 2-4 m above mean tide, have little elevation to spare. Models of melted ice and thermal expansion alone suggest that sea levels will increase by perhaps two meters at the end of the century compared with records from the year 2000. However, even these levels do not include wave-driven flooding from tropical or winter storms. Low islands once protected by reef crests as a breakwater will lose that critical function with higher water levels [46]. Even now, equatorial low islands such as Tarawa experience high tides nearly 1.5 m above sea level and are frequently flooded during modest coincident storms, or worse still, during king tides. In addition, El Niño conditions raise sea levels by another half meter and are an additional source of seawater flooding [45]. During storm events, floods wash through villages causing damage to houses, saltwater intrusion in the freshwater supply, and crop losses, seawalls notwithstanding. Taro is typically grown in the interior of atoll islands, and they are typically planted in pits that are dug to contact the freshwater lens. This makes them vulnerable to salt-water intrusion as a result of storm surges because these pits are the lowest point on land. By contrast, while El Niño brings warmer air and higher rainfall to the central Pacific, La Niña years that often follow produce the opposite effect. Droughts, which can last from 24 to more than 40 months, currently occur in Kiribati every 6–7 years [47]. During these events, drinking and agricultural water supplies are diminished, rainwater tanks are quickly exhausted, and many domestic wells tapping into thin fresh groundwater lenses become saline.

Although population pressures are not quite as severe in the Marshall Islands, the problems of sea-level rise and flooding are essentially the same (Fig. 8.8). In the highly urbanized capital of Majuro, the shoreline facing the windward side has expanded, but as in the case of Tarawa, that change has been largely driven by widespread beach enlargement for a mix of residential, commercial, and industrial activities [48]. However, unlike the nation of Kiribati, the Marshallese have an escape hatch called compact of free association with the United States (Chap. 7), which allows the people of RMI to live and work in the US without a visa. Indeed, a third of their 60,000 citizens now reside in the US. Many have left for economic reasons, but others have been referred to as *climate change refugees*. Of course the Marshall Islands have some unique problems, including Runit Island on Enewetak Atoll where radioactively contaminated soil has been stored in a sea-level concrete crypt (Fig. 7.11). Built in the 1970s to contain an estimated 73,000 cubic meters of contaminated topsoil, the dome is now cracked, deteriorating, and leaking radioactive waste into the groundwater with the rise and fall of the tides [49].

The Special Case of the Solomon Islands

In contrast to sea-level rise on low islands of the central Pacific, increases up to three times the global mean have been observed in the western tropical Pacific by satellite altimetry. This is especially true for the region north of New Guinea and the Solomon Islands where the Pacific Warm Pool resides (Fig. 1.1). This is Earth's largest region of warm sea surface temperatures; it has the highest rainfall, and it is fundamental to global atmospheric circulation. Significantly, the pool has warmed and grown substantially during the past century and it has also experienced the world's highest rates of sea-level rise in recent decades. This in turn has lead to substantial impacts on small island states in the region [50, 51]. Aerial and satellite reconnaissance of the northern Solomon Islands since 1947 has shown considerable shoreline recession in this part of the world. Vegetated reef islands have disappeared or have experienced severe erosion and despite being sparsely populated, entire villages have had to move to higher ground, especially those exposed to the open ocean. Indeed the Solomon Islands have been referred to as a global hotspot for sea-level rise [52].

Fig. 8.8 Residents in the northern part of the capital city of Majuro in the Marshall Islands watch as their neighborhood floods with seawater during a king tide. This high tide followed flooding from storm surge earlier that day (March 3, 2014) (Photo by Karl Fellenius, University of Hawaii Sea Grant)

8.5   Climate Change and Coral Reefs

Corals are colonial, meaning that they are composed of thousands of small tubular animals called polyps, each of which has a mouth surrounded by tentacles. The polyps within a colony are genetically identical and their proliferation results in colonial growth. Each polyp maintains a symbiosis with microscopic single-celled photosynthetic plants called zooxanthellae (where the x is pronounced as a z). The plant cells are typically so numerous within the coral's tissue that they impart a golden brown color to the polyps (Fig. 8.9 insets). Corals depend on their zooxanthellae for nutrition- up to 90% of it- in the form of photosynthetically produced organic matter, the excess of which they leak to their hosts. They also take coral waste in the form of nitrogen and phosphorous as plants typically do, and thereby efficiently recycle nutrients between the partners. Corals build a limestone skeleton around themselves and as they grow, the reef itself is enlarged and strengthened. Zooxanthellae participate in that calcification process so that even a cloudy day can reduce the amount of skeleton produced by 50%. Corals thereby rely intimately on

Fig. 8.9 Bleached corals on the Great Barrier Reef (foreground) with visible white skeletons compared with normally pigmented unbleached corals (background). Corals are composed of tubular polyps with tentacles all of which typically exhibit a golden brown color (*lower* inset) due to their symbiotic algae, shown highly magnified (upper *right*) (Images Wikipedia.org, Allison Lewis (upper *right*) and the National Coral Reef Institute (lower *left*))

their zooxanthellae partners. The only problem is that the symbiotic partnership is tempestuously delicate and requires a number of conditions including a narrow range of temperatures that is typically somewhere between 20°C and 30°C. The actual upper temperature limit may vary slightly depending on a number of factors including geographic location, coral species and the exposure time [53]. In the past, tropical waters rarely exceeded the 30°C limit, at least not for prolonged periods, but that is not the case any more. With global warming, waves of high temperature (only a degree or two above the upper limit) have become more common, and one of the first signs of that thermal stress is the expulsion of zooxanthellae. As expulsion develops over days or weeks the coral tissues become transparent, allowing the white skeleton supporting the coral to become visible. When most of the zooxanthellae are gone the coral is referred to as being 'bleached' (Fig. 8.9).

Bleached corals begin using their energy reserves as their zooxanthellae populations become diminished. Should normal conditions return, the animals will regain their symbionts, but if warm water persists, the corals will slowly starve and become susceptible to disease. Within a month or two, corals deprived of their symbionts will typically die. Other organisms then bore into the skeletons and the entire reef begins to erode and collapse.

Canaries in the Coalmine

Prior to the 1980s only one large-scale bleaching event had ever been recorded. However, every year since then, bleaching on a regional scale has been reported somewhere in the world. The first global bleaching occurred in 1982–83 and corresponded with an extreme El Niño event. At that time it was described as "the most widespread coral bleaching and mortality in recorded history" [54]. That record lasted 14 years.

In 1997–98, a second 'super El Niño' hit the tropics. Reefs in the Pacific from the Great Barrier Reef and the Philippines in the west to Hawaii and French Polynesia in the east were hit hard, although Indian Ocean reefs took the brunt of the warm water. In that region, temperatures of more than 30°C killed more than 90% of all corals in shallow water (to a depth of 15 m) [55]. At the time this was the warmest temperature recorded in the region in more than 200 years. The precise relationship between mass bleaching and El Niño effects outside of the Pacific are unclear, but the two are correlated. Roughly 16% of the world's reefs were destroyed in a matter of months and it was estimated that if recovery takes place at all, it would require 20–50 years [56]. Regrettably, these reefs did not get that much time. In 2010, the same area was hit again by a global bleaching event, killing the recovering and many of the surviving corals. Four years later a far stronger El Niño began in the western Pacific, primarily north of the equator. However, by late 2015–2016 it was still strengthening and extended in a line east of the equator as well as south to the Great Barrier Reef (Fig. 8.10).

Thus began the third and by far the worst global bleaching event thus far since 1983, which elevated water temperatures to more than 31°C. During that time more than 67% of corals in the northern Great Barrier Reef died, including reefs that extend north to Papua New Guinea and the Solomon Islands, and east to include New Caledonia, Fiji and Samoa. A swath of hot water along the equator killed more than 95% of corals in parts of Kiribati, and extended north as severe bleaching was recorded in Hawaii and the Mariana Islands as well [57, 58]. The massive bleaching episode that started in 2014 continued in 2016 and extended into the central part of the Great Barrier Reef that had escaped severe bleaching in the previous year. The combined impact of this back-to-back bleaching stretches for 1500 km, leaving only the southern third unscathed. This global event is now the longest, most widespread and most damaging ever recorded, on one of the best-managed coral reefs in the world [59, 60]. As of mid-2017, bleaching alerts were waning, but persisted in the western and eastern Pacific, as well as in the NW Hawaiian Archipelago.

To put this in perspective, more than 75% of the world's reefs lie in the Pacific Ocean, but roughly 1500 km^2 of them have disappeared annually between 1968 and 2004, and that rate has doubled since 1995 [61, 62]. Assuming no changes in greenhouse gas emission, projections of current rates of reef destruction suggest that by the 2030s roughly half of the reefs in the world will bleach in most years, and by the 2050s that percentage is expected to grow to 95% [62]. It should also be noted that as carbon dioxide increases in the atmosphere, more of it dissolves in seawater. This

Fig. 8.10 Bleaching alerts based on sea surface temperatures and corresponding levels of high probability of coral bleaching, 2014–2016. Highest temperatures are shown in red. A global El Niño coral-bleaching event began in 2014 and became fully developed in 2015–2016. (Courtesy of NOAA Coral Reef Watch [58])

forms carbonic acid, which increases ocean acidity. It also changes the chemistry of seawater in a way that makes it more difficult for calcifying plants and animals (including corals) to build their skeletons. This *ocean acidification*, as it is called, not only reduces coral growth, it also makes their skeletons more porous and brittle. The term 'coral osteoporosis' is applicable, and adds another layer of global threat to the issue of warming oceans. It has been suggested that even with good management, coral cover is likely to decrease from the present-day maximum of 40% to 15–30% by 2035 [61].

8.6   The Marine Realms of Papua New Guinea and Kiribati

Papua New Guinea

PNG's principal marine and coastal ecosystems include 13,840 km² of coral reefs, 4200 km² of mangrove forests, and extensive seagrass beds. There are at least 500 species of stony corals that compose the framework of complex coral reefs in the region including fringing reefs, barrier reefs and atolls. These are especially well developed (and poorly known) around the peninsula-like southeastern part of the country (Fig. 8.3). Reefs in turn provide a home for 1635 reef-associated fish species. Reefs also protect the shoreline from erosion and from physical damage caused by high waves and storm surge, and provide sheltered reef lagoons which serve as protected habitats for seagrass (seven species) and mangroves (34 species). Both of these marine plant groups grow in muddy and sandy environments that become stabilized by their root systems, and thereby form habitat and nursery grounds for thousands of fishes and invertebrate species. Many reef fishes as well tuna and other commercial species of global importance employ mangroves as spawning and juvenile growth areas before migrating offshore.

Mangroves are particularly well developed around river deltas in PNG, although it should be noted that West Papua is home to the largest mangrove forests in the world where they are thought to encompass an area of more than 29,000 km^2. Bintuni Bay in the Bird's Head Peninsula alone (Fig. 8.3) contains 4500 km^2 of these forests [63]. However, mangroves are disappearing because of oil pollution that clogs the pores they use for gas exchange. Mangroves are also extracted for construction, as well as for wood chips and charcoal, particularly in the Gulf of Papua [64]. There are also extensive mangrove forests in the Solomon Islands and in Fiji where similar problems exist. Climate change poses an additional element of uncertainty for these valuable ecosystems. On one hand increases in temperature and CO_2 may increase their productivity and allow them to expand their range. However, at the same time, sea-level rise threatens to bury them with sediment, or force them to retreat inland by inundation stress and increases in salinity [65].

Coastal development, including roads and housing produce runoff that reduce water clarity or blanket nearshore communities with sediment. Loads of nutrients from fertilizer use in agriculture, and untreated sewage from a growing coastal population cause blooms of algae that turn the water green and sometimes toxic. Population growth rates in coastal areas are rising, as is the pace of coastal development and demand for cash income. Indeed it has been suggested that population growth and its associated effects described below are now or soon will be one of the most important drivers of food insecurity at least in some areas of PNG [66]. Increased coastal subsistence and artisanal fishing activity is a significant part of this issue and the sustainability of coastal fish stocks is becoming a concern. Places in New Guinea that were once lightly populated and endowed with a diverse source of reef fish are now near population centers that are both highly stressed and over-fished [67]. This raises the possibility of 'Malthusian overfishing' here as elsewhere, in which immediate food needs and income override long-term sustainability and conservation of resources [68]. As described below, such effects are not limited to nearshore environments or to populations based solely in New Guinea.

The Exclusive Economic Zone and the PNG Tuna Fishery

Papua New Guinea has established a 12 nautical mile territorial sea and a 200 nautical mile Exclusive Economic Zone (EEZ) around its perimeter, as have other countries (Fig. 8.11). Indeed, PNG has one of the largest such zones in the Pacific

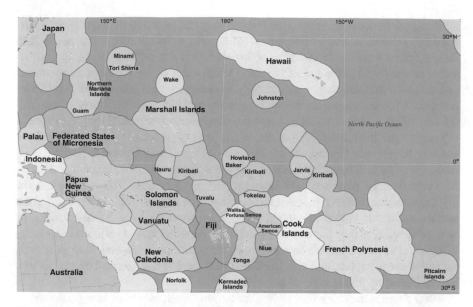

Fig. 8.11 Exclusive Economic Zones of the tropical Pacific. After CartoGIS, College of Asia and the Pacific, The Australian National University

(after French Polynesia and Kiribati), and all 2.4 million square kilometers of it has been declared a controlled fishing zone. More than half of the world's tuna is caught in the western Pacific, and 10–20% of it is taken in Papua New Guinea's EEZ. This is the country's largest and most commercially important fishery, although highly mobile factory ships from China, Japan, Taiwan South Korea, Spain, and the Americas are responsible for most of the catch [69, 70].

The primary species targeted include albacore, skipjack, yellowfin and bigeye; the latter three are tropical species concentrated between 10° north and south of the equator. The following summarizes the state of the tropical Pacific tuna fishery including those taken from PNG waters.

Skipjack are relatively small, surface-schooling species caught by both purse seines and longlines (described below). Most of the catch is sent to canneries. The stocks appear to be sustainably fished at present levels as long as reproduction and juvenile mortality remain as they are. In other words, there is little capacity that remains for any increase in fishing pressure. Yellowfin tuna are fast-growing species that reach 90 cm in the first 18 months and mature at 2–3 years of age. Nonetheless, these tuna appear to be fully exploited by purse seine and longline, meaning that increased fishing pressure will cause a decline in the stock. Indeed, population decreases have led to the recommendation by the Western and Central Pacific Fisheries Commission of limiting catches to 2004 levels. Bigeye tuna are similar in size and appearance to yellowfin, but adults tend to live in deeper, cooler water where they develop a relatively high fat content and command a very high price in the sashimi market. This species is currently overfished across the tropical Pacific and has been in this condition since the late 1990s [71, 72].

Purse seines are often more than 1–2 km long and over 200 m deep. These encircle large schools of tuna, often using helicopters to locate them. The net is then slowly drawn together from below and closed near the surface similar to a drawstring purse. Legal purse seines are fitted with exclusion devices that allow turtles and dolphins that are caught accidentally to escape. Illegal nets typically make no provisions for reducing such 'bycatch', species including sharks, juvenile tuna, and others that are caught in addition to the target species. Commercial longlines consist of one or more main lines, each often several km long, which in the case of tuna are set near the surface. Thousands of shorter lines are suspended from the main line, and each of these is baited with hooks. Longline fishing is indiscriminant and therefore controversial because of the large proportion of bycatch. The same is true for drift nets, which are net curtains a km or more long that are held up by floats at the surface and made taught by weights suspended from the base. Made of nylon or similar synthetics, they are almost invisible in the water and are not biodegradable. They are also indiscriminate in what they catch, and if lost they become ghost nets that continue to capture and kill. High-seas drift netting has declined as a result of a ban instituted by PNG among other nations.

PNG charges licensing fees to foreign fishing fleets. However to regulate and at the same time maximize value from tuna – including much needed jobs and infrastructure – the PNG government now offers long-term fishing licenses to countries that invest in domestic tuna processing plants. However, even with controls on total allowable catches, tuna stocks have declined in part due to illegal, unreported, and unregulated fishing in their waters, especially on the high seas where regulations are difficult to enforce [73].

Marine Protected Areas in Papua New Guinea

One of the ways in which island states can manage their resources is by the creation of Marine Protected Areas. There are many types of MPAs, however, and they may include marine sanctuaries, marine parks, marine reserves, national parks, conservation areas, wildlife management areas, and a variety of other terms that may extend protection to certain resources. Some MPAs protect certain species seasonally, or protect them in certain areas only, and thus an MPA may be open to at least some extractive use including mining and bottom trawling. By contrast, a 'reserve' indicates an area where no fishing or extraction of any type is allowed, and a 'no-take reserve' is an appellation that is sometimes used to ensure that there is no ambiguity. There are 22 marine protected areas in Papua New Guinea [74], but protection is often lacking due to resource deficiencies including adequate staff for monitoring and enforcement. Such MPAs are often referred to as *paper parks*. However, the Nature Conservancy has designed one of the most successful and best-known MPAs in Kimbe Bay, a highly diverse area on the north coast of New Britain (Bismarck Archipelago) that is threatened from a variety of sources including many of those described above. The management plan consists of focusing on local community engagement and education that includes conservation planning and development. Communities in several areas of the bay are implementing plans to sustainably manage their resources and increase their no-take zones, which seem to work especially well in villages within line-of-sight of the areas to be protected [75].

Climate Change and Management of Marine Resources in PNG and Kiribati

The central Pacific island nation of Kiribati encompasses part of the Line Islands to the east, the Gilbert Islands in the west, and the Phoenix Islands in the center (Fig. 1.21). There are 33 islands comprising only 811 km^2, but they are scattered over an exclusive economic zone of 3.5 million km^2, an area more than one third of the continental United States. Kiribati's EEZ is an important tuna fishing zone for industrial fleets from a number of distant-water nations. Like PNG, tuna fishing is an economic mainstay for Kiribati, and the government earns the majority of its revenue from licensing overseas fishing vessels and selling access days to fish in its waters. Although it often does not reveal how many days it sells, fees to fish its fertile waters have more than tripled between 2012 and 2015 [76]. Part of Kiribati's success is due to migration of tuna from Papua New Guinea during El Niño periods. Thus as waters warm in the western Pacific, the areas that are favorable to tuna seining move eastward to Kiribati, resulting in large tuna catches there, especially skipjack and yellowfin. Models of what may happen as Pacific surface waters warm, suggest that the Pacific Warm Pool will become less suitable for tuna feeding and spawning, resulting in a redistribution of tropical tuna toward the east. However, it is also anticipated that tropical tuna are likely to move progressively into cooler subtropical and easterly tropical areas, ultimately reducing their abundance in western EEZ territories [77].

Kiribati's MPA

All 33 of Kiribati's islands are low and are surrounded by fringing or barrier reefs, although fourteen of them are atolls [78]. The Phoenix Islands are essentially uninhabited (fewer than 50 people, all on one atoll) and in 2008 Kiribati designated the region as the Phoenix Islands Protected Area, a marine reserve and a UNESCO World Heritage site. The PIPA site encompasses more than 400,000 km^2, roughly the size of California, and includes eight islands with their reefs, and their waters to a depth of 5000 m. The entire area was closed to commercial fishing in 2015. With few people, little or no fishing and almost no pollution, the area is essentially pristine but unfortunately for its reefs, coral bleaching due to high sea-surface temperatures occurred due to El Niño events in 2002–2003, 2010 and 2015–2016. Relatively rapid recovery took place after the 2010 episode, but the time for full recovery is estimated at 8–15 years and the reefs did not have the luxury of time.

Persistently high ocean temperatures in 2015–2016 killed most of the corals in the region [78]. Thus, with decreasing intervals between warm-water events, there is no safe haven even in the middle of the equatorial Pacific in a marine protected area. This is in agreement with the greenhouse-gas warming of ocean water as the driver of increased El Niño frequencies, particularly in the equatorial Pacific [79].

The Future of Pacific Reefs

Several different types of assaults on coral reefs have been addressed in this brief outline. Some local or regional problems such as poor land management (e.g., nutrients from sewage and agriculture, deforestation and erosion, sediment from construction, toxins and pathogens from storm water runoff) are becoming more common and widespread, as is the problem of population-pressured overfishing. However, none of these can be viewed as independent actors, especially when combined with rising sea levels, global warming and ocean acidification. On one hand, distinct combinations of stressors may have greater effects in some areas than in others. On the other hand, such impacts are likely to be interactive and increase the overall effect of stress on this complex ecosystem [80]. Even so, if an isolated mid-ocean region such as PIPA suffers from repeated and serious bleaching episodes, there is no doubt that reefs worldwide are in deep trouble. Worse still, these effects have evolved from a rise in global temperatures of about 1°C since pre-industrial times. While there is still hope that a reduction in the ten billion tons of carbon we put into the atmosphere last year will decrease, and the inexorable march to more days in hot, acidified water will abate, more CO_2 is expected before any reduction occurs. It is expected that we will add perhaps another degree C of global warming by the end of this century, and that is an optimistic projection. Given this grim outlook, conserving coral reefs as they existed during the last century is no longer a realistic option [81]. Research for the last 20 years has focused on the theme of *resistance,* the ability to withstand environmental change, and *resilience*, the ability to regain the status quo of a health reef system even after stressful events. This work has shown that certain types of corals are more resistant to bleaching than others [82], although the mechanisms are unclear. Certain types of zooxanthellae also appear to be resistant to bleaching and there has been an intense debate concerning the role of these symbionts in that process [83]. In either case, encouraging the conservation of species that resist bleaching, and those that bleach and recover at higher temperatures would be prudent. Reefs in geographic regions with high natural variability in temperature appear to be more resilient than reefs in areas where temperatures are highly constrained and are then suddenly exposed to anomalous heat [84]. Likewise, reefs that are removed from human interference recover more quickly than those that are overfished or subject to the influence of poor land management [85]. Creation of MPAs in such places that are real no-take marine reserves is a step in the right direction, but such places by themselves will not address which species are most important in managing the structure and function of reefs, and steering them through the changes the next century will bring.

References

1. Prism, SPC International (2016) https://prism.spc.int/regional-data-and-tools/population-statistics/169-pacific-island-populations
2. Hansen MC, Potapov PV et al (2013) High-resolution global maps of 21st century forest cover change. Science 432:850–853
3. Mockler SB (1995) Water vapor in the climate system. Am Geophys Union. http://www.eso.org/gen-fac/pubs/astclim/espas/pwv/mockler.html
4. Zimov SA, Schuur EA, Chapin FS (2006) Permafrost and the global carbon budget. Science 312:1612–1613
5. Edwards CR, Trancik JE (2014) Climate impacts of energy technologies depend on emissions timing. Nat Clim Chang 4:347–352
6. Anthony KW, Daanen R et al (2016) Methane emissions proportional to permafrost carbon thawed in Arctic lakes since the 1950s. Nat Geosci 9:679–682
7. Portmann RW, Daniel JS, Ravishankara AR (2012) Stratospheric ozone depletion due to nitrous oxide: influence of other gases. Phil Trans Roy Soc B 367:1256–1264
8. Adams J, Maslin M, Thomas E (1999) Sudden climate transitions during the quaternary. Progr Phys Geogr 23:1–36
9. Hansen JE, Sato M (2012) Paleoclimate implications for human-made climate change. In: Berger A, Mesinger F, Sidjaki S (eds) Climate change: implications for paleoclimate and regional aspects. Springer, Vienna, pp 21–48
10. Ogburn SP (2013) Ice-free Arctic in the Pliocene, last time CO_2 levels above 400 ppm. Nature Magazine May 10, 2013. https://www.scientificamerican.com/article/ice-free-arctic-in-pliocene-last-time-co2-levels-above-400ppm/
11. Raymo ME, Mitrovica JX et al (2011) Departures from eustasy in Pliocene sea-level records. Nat Geosci 4:328–332
12. Nunn PD (2007) Climate, environment and society in the Pacific during the last millennium. Elsevier, Oxford
13. Mann ME, Zhang Z et al (2009) Global signatures and dynamical origins of the little ice age and the medieval climate anomaly. Science 326:1256–1260
14. Comiso J (2012) Large decadal decline of the Arctic multiyear ice cover. J Clim 25:1176–1193
15. Rosen J (2017) After the ice goes. Nature 542:152–154
16. Zwally HJ, Li J et al (2015) Mass gains of the Antarctic ice sheet exceed losses. J Glaciol 61:1019–1036
17. Rintoul SR, Silvano A et al (2016) Ocean heat drives rapid basal melt of the Totten Ice Shelf. Sci Adv 2:e1601610. https://doi.org/10.1126/sciadv.1601610
18. Zemp M, Gartner-Roer I et al (2015) Historically unprecedented global glacier decline in the early 21st century. J Glaciol 61:745–762
19. Sweet W, Park J et al (2014) Sea-level rise and nuisance flood frequency changes around the United States. NOAA Technical Report NOS CO-OPS 073
20. Yin J, Schlesinger ME, Stouffer RJ (2009) Model projections of rapid sea-level rise on the northeast coast of the United States. Nat Geosci 2:262–266
21. Wdowinski S, Bray R et al (2016) Increasing flooding hazard in coastal communities due to rise in sea level: case study of Miami Beach. Ocean Coast Manag 126:1–8
22. Climate Change Impacts in the United States (2014) National climate assessment. U.S. Global Change Research, Washington, DC. http://nca2014.globalchange.gov/highlights/overview/overview
23. Goldberg W (2016) Atolls of the world: revisiting the original checklist. Atoll Res Bull #610
24. Clements JF (2000) Birds of the world: a checklist. Cornell University Press, Ithaca
25. Sherman PL, Ash J et al (2009) Forest conversion and degradation in Papua New Guinea 1972–2002. Biotropica 41:379–390
26. Nelson PN, Gabriel J et al (2014) Oil palm and deforestation in Papua New Guinea. Conserv Lett 7:188–195

27. Tolia DH, Petterson MG (2005) The Gold Ridge Mine, Guadalcanal, Solomon Islands' first gold mine: a case study in stakeholder consultation. Geol Soc Spec Pub 250:149–159
28. Banks G (2013) Mining. In: Rappaport M (ed) The Pacific islands: environment and society. University of Hawaii Press, Honolulu, pp 379–391
29. Kirsch S (2008) Social relations and the green critique of capitalism in Melanesia. Am Anthropol 110:288–298
30. Papua New Guinea Climate variability, extremes and change in the Western Tropical Pacific: new science and updated country reports 2014. http://www.pacificclimatechangescience.org/wp-content/uploads/2014/07/PACCSAP_CountryReports2014_WEB_140710.pdf
31. Corely RHV, Tinker PBH (2016) The oil palm, 5th edn. Wiley Blackwell, Hoboken
32. Lang ALS, Omena M et al (2015) Climate change in Papua New Guinea: impact on disease dynamics. Papua New Guinea. Med J 58:1–10
33. Leong J-A, Marra JJ et al (2014) Hawai'i and U.S. affiliated Pacific islands. Climate change in the United States, ch. 23. In: Melillo JM, Richmond TC, Yohe GW (eds) U.S. Global Change Research Program, pp 537–556. http://nca2014.globalchange.gov/report/regions/hawaii-and-pacific-islands#intro-section-2
34. Liao W, Atkinson CT et al (2017) Mitigating future avian malaria threats to Hawaiian forest birds from climate change. PLoS One 12(1):e0168880. https://doi.org/10.1371/journal.pone.0168880
35. Benning TL, Pointe L et al (2002) Interactions of climate change with biological invasions and land use in Hawaiian Islands: modeling the fate of endemic birds using a geographic information system. Proc Natl Acad Sci USA 99:14246–14249
36. Atkinson C (2005) Ecology and diagnosis of introduced avian malaria in Hawaiian forest birds. https://pubs.usgs.gov/fs/2005/3151/report.pdf
37. http://www.botany.hawaii.edu/basch/uhnpscesu/pdfs/NatHistGuideAS09op.pdf
38. Burns WCG (2002) Pacific island developing country water resources and climate change. Ch. 5. In: Gleick P (ed) The world's water: 2002–2003 the biennial report on freshwater resources. The Pacific Institute, Oakland, pp 113–131
39. White I, Falkland T et al (2008) Safe water for people in low, small Island Pacific Nations: the rural-urban dilemma. Development (Cambridge) 51:282–287
40. Woodroffe CD (2008) Reef-island topography and the vulnerability of atolls to sea-level rise. Glob Planet Chang 62:77–96
41. Web AP, Kench PS (2010) The dynamic response of reef islands to sea-level rise: evidence from multi-decadal analysis of island change in the Central Pacific. Glob Planet Chang 72:234–246
42. Kench PS, Thompson D et al (2015) Coral islands defy sea-level rise over the past century: records from a Central Pacific atoll. Geology 43:515–518
43. Biribo N, Woodroffe C (2013) Historical area and shoreline change of reef islands around Tarawa, Kiribati. Sustain Sci 8:345–362
44. Duvat V (2013) Coastal protection structures on Tarawa Atoll, Republic of Kiribati. Sustain Sci 8:363–379
45. Impact of climate change, the low islands, Tarawa Atoll, Kiribati (2000) In: Bettencourt S, Warrick R (eds) Cities, sea and storms. Managing resources in Pacific Island economies, vol 4. The World Bank, Washington, DC, p 19–26. http://siteresources.worldbank.org/INTPACIFICISLANDS/Resources/4-VolumeIV+Full.pdf
46. Storlazzi CD, Edwin PL, Berkowitz E, Berkowitz P (2015) Many atolls may be uninhabitable within decades due to climate change. Sci Rep 5:14546
47. White I, Falkland T et al (2007) Climatic and human influences on groundwater in low atolls. Vadose Zone J 6:581–590
48. Ford M (2012) Shoreline changes on an urban atoll in the central Pacific Ocean: Majuro Atoll, Marshall Islands. J Coast Res 28:11–22
49. http://www.washingtonpost.com/sf/national/2015/11/27/a-ground-zero-forgotten/

50. Palanisamy H, Meyssignac B et al (2015) Is anthropogenic sea-level fingerprint already detectable in the Pacific Ocean? Environ Res Lett 10:084024. https://doi. org/10.1088/1748-9326/10/8/084024
51. Weller E, Min S-K et al (2016) Human-caused Indo-Pacific warm pool expansion. Sci Adv 2(7):e1501719. https://doi.org/10.1126/sciadv.1501719
52. Albert S, Leon JX et al (2016) Interactions between sea-level rise and wave exposure on reef island dynamics in the Solomon Islands. Environ Res Lett 11:054011. https://doi. org/10.1088/1748-9326/11/5/054011
53. Jokiel PL (2004) Temperature stress and coral bleaching. In: Rosenberg E, Loya Y (eds) Coral health and disease. Springer, Berlin, pp 401–425
54. Glynn PW (1990) Coral mortality and disturbances to coral reefs in the eastern Pacific. Elsevier Oceanogr Ser 52:55–126
55. Sheppard RC (2003) Predicted recurrence of mass coral mortality in the Indian Ocean. Nature 425:294–297
56. Wilkinson C (2000) Status of coral reefs of the world. Australian Institute of Marine Science, Townsville
57. http://www.noaa.gov/media-release/el-ni-o-prolongs-longest-global-coral-bleaching-event
58. https://coralreefwatch.noaa.gov/satellite/analyses_guidance/global_coral_bleaching_20-17_status.php
59. Biello D (2007) Coral reefs losing ground throughout the Pacific. Sci Am. August 2007
60. Hughes TP, Kerry JT et al (2017) Global warming and recurrent mass bleaching of corals. Nature 543:373–377
61. Bruno JF, Selig ER (2007) Regional extent of coral cover in the Indo-Pacific: timing, extent and regional comparisons. PLoS One 2(8):e711. https://doi.org/10.1371/journal.pone.0000711
62. Burke L, Reytar K et al (2011) Reefs at risk revisited. World Resources Institute, Washington, DC.   http://www.wri.org/sites/default/files/pdf/reefs_at_risk_revisited_executive_summary. pdf
63. Mangubhai S, Erdmann MV et al (2012) Papuan Bird's Head Seascape: emerging threats and challenges in the global center of marine biodiversity. Mar Pollut Bull 64:2279–2295
64. Maclean J, Mallery L (2014) State of the Coral Triangle: Papua New Guinea. Asian Development Bank. https://www.adb.org/sites/default/files/publication/42413/state-coral-tri-angle-papua-new-guinea.pdf
65. Ellison J (2001) Possible impacts of sea-level rise on south Pacific mangroves. In: Noye BJ, Grzechnik MP (eds) Sea-level changes and their effects, Singapore, World Scientific Publishing, pp 49–72
66. Butler JRA, Skewes T et al (2014) Stakeholder perceptions of ecosystem service declines in Milne Bay, Papua New Guinea: is human population a more critical driver than climate change? Mar Policy 46:1–13
67. Drew JA, Amatangelo KL, Hufbauer RA (2015) Quantifying the human impacts on Papua New Guinea reef fish communities across space and time. PLoS One 10(10):e0140682. https:// doi.org/10.1371/journal.pone.0140682
68. Worm B, Branch TA (2012) The future of fish. Trends Ecol Evol 27:594–599
69. Doulman DJ, Wright A (1983) Recent developments in Papua New Guinea's tuna fishery. Mar Fish Rev 45:47–59
70. http://www.fisheries.gov.pg/FisheriesIndustry/TunaFishery/tabid/104/Default.aspx
71. Bailey M, Sumaila UR, Martell SJD (2013) Can cooperative management of tuna fisheries in the Western Pacific solve the growth overfishing problem? Strat Behav Environ 3:31–66
72. Hampton J (2010) Tuna fisheries status and management in the western and central Pacific Ocean.   http://awsassets.panda.org/downloads/background_paper___status_and_manage-ment_of_tuna_in_the_wcpfc.Pdf
73. Havice E, Reed K (2012) Fishing for development? Tuna resource access and industrial change in Papua New Guinea. J Agrar Chang 12:413–435

74. Wood LJ (2007) MPA Global: a database of the world's marine protected areas. http://www.mpaglobal.org/index.php
75. Greene A, Smith SE et al (2009) Designing a resilient network of marine protected areas for Kimbe Bay, Papua New Guinea. Oryx Int J Conserv 43:488–498
76. http://www.reuters.com/article/us-climatechange-elnino-tuna-idUSKBN0TM0F320151203
77. Bell JD, Ganachaud A et al (2013) Mixed responses of tropical Pacific fisheries and aquaculture to climate change. Nat Clim Chang 3:591–599
78. Obura D, Donner SD et al (2016) Phoenix Islands Protected Area climate change vulnerability assessment and management, report to the New England. Aquarium, Boston. http://www.phoenixislands.org/pdf/PIPA-CC-scoping-study-Jan-18-2016.pdf
79. Cai W, Borlace S et al (2014) Increasing frequency of extreme El Niño events due to greenhouse warming. Nat Clim Chang 4:111–116. https://doi.org/10.1038/nclimate2100
80. Ateweberhan M, Feary DA et al (2013) Climate change impacts on coral reefs: synergies with local effects, possibilities for acclimation, and management implications. Mar Poll Bull 74:526–539
81. Hughes TP, Barnes ML et al (2017) Coral reefs in the anthropocene. Nature 546:82–90
82. Fitt WK, Gates RD et al (2009) Response of two species of Indo-Pacific corals, *Porites cylindrica* and *Stylophora pistillata*, to short-term thermal stress: the host does matter in determining the tolerance of corals to bleaching. J Exp Mar Biol Ecol 373:102–110
83. Howells EJ, Beltran VH et al (2012) Coral thermal tolerance shaped by local adaptation of photosymbionts. Nat Clim Change 2.2(2012):116–120. https://doi.org/10.1038/NCLIMATE1330
84. Carilli J, Donner SD et al (2012) Historical temperature variability affects coral response to heat stress. PLoS One 7(3):e34418. https://doi.org/10.1371/journal.pone.0034418
85. Carilli JE, Norris RD et al (2009) Local stressors reduce coral resilience to bleaching. PLoS One 4(7):e6324. https://doi.org/10.1371/journal.pone.0006324

Index

A
Abemama Atoll, 36, 37
Achromatopsia, 35
Age of Discovery, 57
Alteration of Midway, 161–164, 166
American business, 125–127
American Civil War, 126
American Samoa, 10, 20, 33, 37, 70, 143, 158
Anchoveta
    American Guano Company, 139
    description, 133
    dramatic regression of, 148
    enormous schools, 134
    plankton and seabirds, 134
Anglican Church, 120, 131
Antofagasta Company, 149
Arboral Snakes, 105–107
Arquebuses, 64, 71, 72
Arrowroot, 85
Artocarpus altilus, 81
Atoll evolution, 9, 22–26
Atoll form, 20
Atoll reefs, 162
Australia's Great Barrier Reef, 17
Australian Plate, 11, 30
Australian sandalwood traders, 123
Austronesian, 41, 43, 49, 50, 52, 53

B
Banaba, 10, 151, 152, 192
Bananas, 89
Barrier reefs, 9, 15, 17, 23, 24, 199, 203
Bayonet Constitution, 129

Bêche-de-Mer, 99–101
Betio Island, 162, 194
Bible, 120
Bigman, 79
Black rats, 105
Blackbirding, 126, 127
Black-lipped oysters, 103
Bolivia and the Nitrogen Wars, 149, 150
Bonnet-shaped Arno Atoll, 20
Box-like Rose Atoll, 20
Bravo test crater, 170
Breadfruit, 81, 82
    export, 95–96
British Crown Colony, 127
British nuclear testing, 173–175
British Royal Society, 26
Brown tree snake, 106
Butchering, 97

C
Cannibalism, 119
Carbon dioxide and greenhouse gasses, 183
Castle Bravo, 170, 171
Casuarina equisetifolia, 16
Catholic missionaries, 121, 122
Chamorro, 52, 53, 63
Chemical and radiochemical waste, 165
Chile
    borders of, 147, 149
    British-Chilean consortium, 149
    tax and border dispute, 147
Chincha Islands, 134–137, 146
Christmas Island on Christmas Eve 1768, 74

© Springer International Publishing AG 2018
W.M. Goldberg, *The Geography, Nature and History of the Tropical Pacific
and its Islands*, World Regional Geography Book Series,
https://doi.org/10.1007/978-3-319-69532-7

Printed in the United States
By Bookmasters